Jordanus de Nemore:
De numeris datis

Published under
the auspices of the
Center for Medieval
and Renaissance Studies
University of California, Los Angeles

Publications of the
Center for Medieval and
Renaissance Studies, UCLA

A CRITICAL EDITION
AND TRANSLATION

BY Barnabas Bernard
Hughes, O.F.M.

Jordanus de Nemore

De numeris datis

University of California Press
Berkeley · Los Angeles · London

The emblem of the Center
for Medieval and Renaissance Studies
reproduces the imperial eagle
of the gold *augustalis* struck
after 1231 by Emperor Frederick II;
Elvira and Vladimir Clain-Stefanelli,
The Beauty and Lore of Coins, Currency and Medals
(Croton-on-Hudson, 1974), fig. 130 and p. 106.

University of California Press
Berkeley and Los Angeles, California

University of California Press, Ltd.,
London, England

1   2   3   4   5   6   7   8   9

Library of Congress Cataloging in Publication Data

Jordanus Nemorarius, fl. 1230.
De numeris datis.

(Publications of the Center for Medieval and
Renaissance Studies, UCLA; 13)
A revision of the editor's thesis—Stanford University, 1970.
Bibliography: p.197
Includes index.
1. Algebra—Early works to 1800.   I. Hughes
Barnabas.   II. Title.   III. Series: California.
University. University at Los Angeles. Center for
Medieval and Renaissance Studies. Publications; 13.
QA32.J6713   512   80-21719
ISBN 0-520-04283-2

To Dee and Frank Castanier

Contents

Preface

My high school teacher, Father Francis Guest, O.F.M., once remarked, "If you want to understand anything well, study its history." When I began to teach high school mathematics, his advice prompted me to study the history of mathematics, particularly the works on which my courses were based. An early find was Louis Karpinski's English translation of Robert of Chester's Latin translation of al-Khwārizmī's *Kitāb al-jabr wa'l muqābala*. Sometime later I began to wonder about an early advanced algebra: had the Latin West produced nothing in the Middle Ages?

The comments of older historians of mathematics (Archibald, Bell, Cajori, Smith—to mention a few) suggested that the Middle Ages were mathematically insignificant. On the contrary, as I discovered, the Middle Ages abounded with mathematicians and their works. Witness, for instance, the productive efforts of the twelfth-century translators, such as Campanus of Novara and John of Seville, authors in their own right. The thirteenth century produced Jordanus de Nemore, Leonardo da Pisa, John of Tinemue, John Peckham, and Sacrobosco. From the fourteenth century, there is much to be learned from the Four Calculators, Nicole Oresme, and Giovanni di Casali. These men, at least, deserve an advocate to correct the misstatements of respected writers.

Such antecedents led me toward this present work. I chanced on Jordanus' *De numeris datis*, the advanced algebra that complements al-Khwārizmī's work. Further investigation produced my doctoral dissertation, *The De numeris datis of Jordanus de Nemore: A Critical Edition*,

Analysis, Evaluation and Translation (Stanford University, 1970), of which this book is a revision.

The critical edition has been improved in two ways. First, three additional MSS discovered by my friend and colleague, Dr. Ron B. Thomson, shed new light on the text. Second, a reevaluation of all the MSS now at hand prompted a sifting of the material in the apparatus to remove what I judged to be trivial. The symbolic translation into contemporary algebra required correction and reorganization, and for the lexicographer I have added a glossary. The bibliography has been brought up to date. Finally, the introduction was revised and an index added.

No work is brought to press without the generous assistance of many people. I respectfully acknowledge the patient encouragement of my dissertation chairman, Menahan M. Schiffer, and advisor, Alan Bernstein (both of Stanford University), my editors, Richard Rouse (UCLA Center for Medieval and Renaissance Studies) and Abigail Bok (University of California Press), Marshall Clagett (Institute for Advanced Studies), Michael S. Mahoney (Princeton University), George Molland (University of Aberdeen), Ron B. Thomson (Pontifical Institute of Medieval Studies), John B. Hancock (California State University, Hayward), and Reverend Dr. Jacek Przygoda.

From these libraries I received microfilm and information about manuscripts: Basel, Oeffentliche Bibliothek (Max Burckhardt, Martin Steinmann); Cambridge University Library (H. L. Pink, A. E. B. Owen); Dresden, Sächsische Landesbibliothek (Helmut Deckert, Burghard Burgemeister, W. Stein); Firenze, Biblioteca Nationale (Dr. Auloufi); Göttingen Universität-Bibliothek (Dr. Haenel); Krakow, Biblioteka Jagiellonska (Władyłsaw Serczyk, Jan Pirozynski); Leipzig, Universitäts-Bibliothek; Milano, Biblioteca Ambrosiana (Angelo Paredi); London Science Museum Library (S. A. Jaywardene); Columbia University Library; Bodleian Library (Bruce Barker-Benfield); Paris, Bibliothèque Mazarine (D. Masson, Pierre Gasnault); Bibliothèque Nationale (Denise Bloch); Saint Louis, Vatican Film Library (Charles Ermatinger); Urbana, University Library (Marcella Grendler); Vatican City, Biblioteca Apostolica (Annaliese Maier); and Vienna, Nationalbibliothek (Otto Mazel, Eva Inblich).

Also helpful were my readers; the Shell Companies Foundation, which provided the funds for my doctoral work; and finally, my religious superiors in the Franciscan Order, who permitted me the time to pursue study at Stanford. Of considerable importance was the W. Fenlon Nicholson Fund, which supplied necessary financial assistance toward the completion of this work. The dedication recognizes more than yeoman work of close friends who made much of this possible.

To all, my sincere thanks.

Introduction

He followed not the synthetic but
the analytic way of teaching.
MARINUS on Euclid's *Data*

The *De numeris datis* of Jordanus de Nemore is recognized as the first advanced algebra composed in western Europe. The text assumes the reader's familiarity with fundamental algebraic concepts and skills, and offers a development of quadratic, simultaneous, and proportional equations, for the most part previously unexpressed. Works were available at the beginning of the late twelfth century to provide the foundation. Through them the student became familiar with equations both simple and quadratic, rules for multiplying (what we call) positive and negative integers, monomials and binomials, extraction of roots, and the use of parameters, false position, and the rule of three. Expertise in the fundamentals of algebra signaled a need for a development that would be more abstract and profound. Jordanus provided it.

The Man and His Work

Although the authenticity of his work has been established,[1] biographical information about Jordanus is sparse.[2] Since no mention of his name has been found in any list of clerics, it is supposed that he was a layman. At the earliest, Jordanus flourished in the late twelfth century: his work *De numeris datis* presumes that its readers were familiar with elementary algebra, and this knowledge did not break upon Europe before

Robert of Chester's translation of al-Khwārizmī's *Liber algebre*, in 1145. At the latest, Jordanus completed most if not all of his writing by the mid-thirteenth century. The name "Jordanus de Nemore" is mentioned four times in the *Biblionomia* of Richard de Fournival (circa 1250).[3] Hence, it seems reasonable to assume that Jordanus lived during the last part of the twelfth and the first part of the thirteenth centuries.

It is more difficult to discover where he worked. A supposed clue put him at the University of Toulouse for a series of lectures. In a manuscript reportedly attributed to Jordanus there is the marginal note, "This is enough to say for the instruction of the students at Toulouse." Ron Thomson has shown, however, that the text does not match any known treatise of Jordanus.[4] Lacking any other evidence, we can conclude only that Jordanus taught in Europe.

Further research regarding his identity has unearthed nothing new, though it has cleared away much speculation. The most serious piece of evidence was based on a remark of the Dominican friar Nicolas Trivet (1258–1328). Sometime professor at Oxford and Paris, he composed a chronology for the period 1136 to 1307. In it he wrote of Jordanus de Saxonia (?–1237), the first successor to Saint Dominic as Master General of the Order of Preachers (commonly called Dominicans): "he was renowed in Paris for his secular knowledge particularly in mathematics; and, as it is said, wrote two very useful books, *Weights* and *Given Lines*."[5] Published attention was first given to this statement by Treutlein in 1879.[6] Subsequently the identification of Jordanus de Nemore with Jordanus de Saxonia was accepted, despite the incorrect title for *De numeris datis*, by Curtze,[7] Chevalier,[8] Cantor,[9] and in our own century by Arons[10] and Schreider.[11] The last two were apparently unfamiliar with the devastating attack on the identification by the Dominican historian Heinrich Denifle.[12] He observed that the oldest chronicle of the Dominican Order (fourteenth century) mentions that Jordanus de Saxonia was an artist (he had taken the liberal arts course) and a theologian, and that it ascribes no mathematical work to him. Furthermore, Denifle found Jordanus de Saxonia nowhere called "Nemorarius." An argument of silence is advanced by Moody and Clagett, who point out that "Jordanus de Saxonia" is not mentioned in any scientific document, while "Jordanus de Nemore" does not appear in any ecclesiastical document.[13] In short, the two men were apparently confused because of similarity of names and academic reputation. Our Jordanus must therefore be a medieval Melchisedec, *sine patre*, *sine matre*, *sine genealogia*, until firsthand evidence to the contrary is found.

Fortunately, we know much more about his works. The attention increasingly directed at them during the past century[14] does credit to a person who is otherwise obscure. Jordanus de Nemore was, and is, recognized as one of the most prestigious natural philosophers of the thirteenth century. His activities encompassed the field of mathematical physics. In particular, he laid the foundation for the entire area of medieval statics.[15] At a more elementary level, his mathematical works on arithmetic, both logistic and specious, and algebra were copied and printed many times, well into the sixteenth century. Only his mathematical works will be considered here.

All the works of Jordanus include abundant mathematics. To narrow the focus of this discussion, only those treatises that are strictly mathematical are identified here. There are six.[16] The *Demonstratio de algorismo* is a practical explanation of the Arabic number system with respect to integers and their use. Similarly, *Demonstratio de minutiis* treats fractions. His *De elementis arismetice artis* became the standard source for theoretical arithemetic in the Middle Ages. The *Liber phylotegni de triangulis* shows off medieval Latin geometry at its best, particularly by giving rigorous geometric proofs of theorems. *Demonstratio de plana spera* is a treatise of five multipartite propositions clarifying various aspects of stereographic projection. Finally, there is *De numeris datis*, the first advanced algebra to be written in Europe after Diophantus (circa A.D. 50).

The absence of an elementary algebra from the list may be justified *a pari*, by considering possible reasons for his composing elementary arithmetics. Both computational and theoretical arithmetic tracts, at a low level of difficulty, were at hand. Jordanus' own works, the *Demonstratio de algorismo* and *de minutiis*, suggest that he was dissatisfied with the content of presentation of the others, just as even today professors produce texts on subjects for which there is already an abundance of books. A side-by-side comparison of Alexander de Villa Dei's *Carmen de algorismo* with the *De algorismo* of Jordanus shows the obvious superiority of Jordanus' later work. Now, there is no firm evidence that Jordanus wrote an elementary algebra.[17] I conjecture that he found no need to do so: an inspection of the two most useful works already at hand, al-Khwārizmī's *Liber algebre* and Fibonacci's *Liber abaci*, strongly suggests that either of these would serve well as introductions. Both texts begin simply with a few definitions and these equations: $x^2 = bx$, $x^2 = c$, and $bx = c$. Quick progress is made to the final equations:

$$x^2 + bx = c, \; x^2 + c = bx, \text{ and } bx + c = x^2.$$

(All these equations, of course, were expressed in words.) Thereafter, much of what is treated today in first-year algebra is covered. Hence, there was no apparent need for Jordanus to compose an elementary algebra.

What purpose would an advanced algebra serve? Each discipline has its elementary and advanced stages. Euclid's geometry introduced the student to the elementary level of the subject. Once equipped with the fundamentals of geometry, the ambitious student was challenged with advanced theorems and nonstandard problems, the one to prove and the other to solve by the method of analysis. Among the analytic books for advanced geometry was Euclid's *Data*, which showed the student how, given certain relationships between lines, planes, and solids, certain other relationships or quantities were thereby found (or given). Seven hundred years ago, no corresponding book existed for the student who would pursue algebra beyond the elementary stages of al-Khwārizmī's work. Jordanus provided one in *De numeris datis*. This text, like Euclid's *Data*, showed the student how, given certain relationships between numbers, certain other relationships or quantities were thereby found (or given).

Analysis and *De datis*

The significance of *De datis* is best appreciated in the context of mathematical analysis. Analysis finds its origin among the Greek geometers. It was thought to have been rediscovered during the Renaissance, but in fact it made an earlier appearance during the later Middle Ages.

The *concept* of analysis was formalized centuries after analysis had become the familiar tool of the mathematician. Mahoney describes in suspenseful fashion the use to which Archimedes puts analysis as "he chases out the consequences of the mathematical situation."[18] It would be half a millennium before mathematicians would gain sufficient distance from the chase for Pappus (circa A.D. 320) to gather together conceptual works on analysis in an essay titled "Treasury of Analysis."

Pappus begins his essay with the caution that it is written for those who have mastered the *Elements* of Euclid and wish to solve more challenging problems. Then he defines analysis. "Now analysis is the passage from the thing sought, as if it were admited, through the things which follow in order (from it), to something admitted as a result of synthesis. . . . If the thing admitted is possible and obtainable, what they call in mathematics given, the (problem) set forth also will be possible, and again the proof will be the reverse of analysis."[19] Logically, Pappus is asserting a biconditional link between what is given and the conclusion one seeks to prove.

He describes a method (analysis) for finding the steps in the proof. Begin with the conclusion, he writes; then, attempt to argue logically "back" to what is given or constructible. If this last stage is reached, then you have—in reverse order—the steps of the proof. Subsequently, one need only reverse these steps, and the truth or solution is established, Q.E.D. or Q.E.F. The bulk of Pappus' essay is devoted to a discussion of the works of various mathematicians, which are useful for perfecting one's analytic ability. Among these works is Euclid's *Data*. There was no counterpart in medieval algebra until Jordanus wrote *De numeris datis*.

What Jordanus accomplished is seen better, I believe, from the viewpoint established by his successor, François Viète. In 1591 Viète wrote *Introduction to the Analytical Art*.[20] His initial chapter contains a résumé of the past: "In mathematics there is a certain way of seeking truth, a way which Plato is said first to have discovered, and which was called 'analysis' by Theon and was defined by him as 'taking the things as granted and proceeding by means of what follows to a truth that is uncontested . . .' The analytic art . . . may be defined as the art of right finding in mathematics."[21] Analysis was apparently reborn. More important Viète was writing about algebra. Indeed, algebra had become analysis.

Three steps are necessary to solve a generalized problem of algebra in an analytic manner. The first two are properly analytic; the third is synthetic and the equivalent of a geometric solution.[22] They are: (1) the construction of the equation; (2) the transformations to which it is subjected until it has acquired a canonical form that immediately supplies the "indeterminate" solution; and (3) the numerical exploitation of the last, i.e., the computation of unequivocally determinate numbers that fulfill the conditions set for the problem. The first two steps require symbols, both numerical and operational. The third step is an example that not only affirms but also clarifies what has preceded. In other words, (1) is the statement of a proposition, (2) lists the directions for working out the statement in general terms, and (3) is a concrete example. This was Viète's approach to algebra.

The algebraic analysis of Viète paralleled that of Pappus. Viète's *Introduction* may well be described as a special body of material prepared for those who wanted, after mastering common calculations, to find the solutions to problems involving general numbers as well as practical problems. For this alone it has been established as useful. Viète certainly introduced the renaissance mathematicans to analysis, but Jordanus gave *De numeris datis* a similar role three hundred fifty years earlier.

Specifically, Jordanus offered his readers an advanced tract on algebraic analysis, analysis in the sense defined by Pappus and restated by Viète. In *De datis* the problem-solver may find which number or ratio relations have solutions. If, in the course of solving a numerical problem, he finds that certain numbers, number relations, or ratios are known, then—assuming his familiarity with the text—he has the solution of his problem immediately at hand. The claim is that *De datis* is a tool chest for numerical analysis.

Let us consider the evidence, measuring Jordanus' work by tripartite norm which Klein defined and applied to Viète's work. A typical theorem or proposition from *De datis* is selected, refashioned in modern symbols, and evaluated. From Book IV, proposition 6:

If the ratio of two numbers together with the sum of their squares is known, then each of them is known.

$$x:y = a, \ x^2 + y^2 = b \qquad (1)$$

Let the ratio of x and y be given. Let d be the square of x and c the square of y: and let $d + c$ be known.

$$x:y = a, \ x^2 = d, \ y^2 = c$$
$$d + c = b \qquad (2)$$

Now the ratio of d to c is the square of the ratio of x and y. Hence; the former is known. Consequently, d and c are known.

$$d:c = x^2:y^2 = a^2 \qquad (3)$$
$$(d/c + 1)y^2 = b \qquad (4)$$
$$y = [b/(a^2 + 1)]^{\frac{1}{2}} \qquad (5)$$

For example, let the ratio of two numbers be 2 and the sum of their squares be 500. Now, since the square of one number is 4 times the square of the other, it follows that 500 is 5 times the square of the other, which makes it 100. The root of this is 10 for the smaller number, and for the larger, 20.

$$y = [500/(4 + 1)]^{\frac{1}{2}} = 10 \qquad (6)$$
$$x = 20$$

In my symbolic translation above, it is clear that (1) is the construction of the equation: the formation of the problem in terms of what is known, a and b, and what is to be found, x and y. Steps (2) through (4) are the transformations to which (1) is subjected until a canonical form, (5), is reached. The final stage, (6), is the numerical exploitation of (5); that is, the computation of unequivocally determinate numbers that fulfill the conditions set for the problem.

This proposition and the others in *De numeris datis* are exercises in problematic analysis after the Greek fashion. As Euclid in his *Data*, so Jordanus in *De datis* assembled those number relations in the form most useful for the working mathematician. Both texts were arranged so that the natural philosopher might refer to them to determine analytically the solution or the solvability of a problem in number relations. Hence, *De datis* was written for both applied mathematics and algebraic analysis.

How could *De datis* have been used? Since any answer is hypothetical, I offer a single example. Jordanus wrote the following in *De ratione ponderis* (R1.06): "If the arms of a balance are proportional to the weights suspended, in such manner that the heavier weight is suspended from the shorter arm, the weights will have equal positional gravity."[23] This is the familiar theorem: Unequal weights in equilibrium are inversely proportional to their distances from the fulcrum. Jordanus established the theorem indirectly by utilizing the principle of work: What suffices to move one weight so many units suffices to move another equal weight the same number of units.[24] The practical problem is that, given two unequal weights and the length of the balance bar, where does one place the fulcrum? *De datis* offers an immediate solution in proposition 6 of Book II: "If the ratio of the two parts of a given number is known, then each of them can be found."

In modern symbols, given $x + y = a$ and $x:y = b$, then both x and y may be determined. Hence, let the given weights, w_1 and w_2, form a known ratio, which by force of the theorem equals the inverse ratio of their distances x and y from the fulcrum:

$$w_1 : w_2 = b = y:x.$$

By appropriate transformations of the two given equations, Jordanus obtained a canonical form that exhibits an indeterminate solution, $y = a/(b + 1)$. Proper substitution and computation provide the respective distances and thus the location of the fulcrum. Jordanus could have used his own algebra in this way to round out a theoretical proof with a practical application.

That *De datis* was used, or at least studied, there seems little doubt. Seven copies of the tract remain from the thirteenth century, six from the fourteenth and fifteenth centuries and two from the sixteenth century, as well as four different digests and revisions. *De datis* caught the imagination of scholars and teachers, as both marginalia and digests indicate. In fact, it received explicit attention from Johannes Widmann von Eger (circa 1489), who lectured at the University of Leipzig on the subject of algebra;[25] and shortly before this, Regiomontanus (1436–1476) intended to publish *De datis*, a plan aborted by his untimely death.[26]

Why *De datis* did not enjoy the same popularity as did for example, analytic geometry or the calculus is understandable: the times were not right. Late medieval scholars may have appreciated its theoretic power, but apparently they did not use it to render their theory more applicable. I think Jordanus tells us why in the prologue to the tract *De ponderibus*: "Since the science of weights is subalternate both to geometry and to natural philosophy, certain things in this science need to be proved in a philosophical manner, and certain things in a geometric manner."[27] He describes the mental framework of the times: qualitative answers are expected from natural philosophy; quantitative answers are drawn from geometry. The cycle is complete in an established paradigm.

The compelling inertia of a paradigm is well described by Kuhn: "In learning a paradigm, the scientist acquires theory, methods, and standards together, usually in an inextricable mixture."[28] Hence, "men whose research is based on shared paradigms are committed to the same rules and procedures for scientific practice."[29] The most obvious confirmation of the *mos geometricus*[30] is found in the philosophical and geometrical demonstrations broadcast in the works of Jordanus, Gerard of Brussels, and the Four Calculators, to mention but a few. The point of view of the natural philosophers ignored any method other than the geometric, any standard other than a physical application. The geometric method was perfectly adequate to trace the proof of a proposition. Natural philosophers were secure in this approach; it had not failed them in their normal problem-solving activity. Practical applications could be roughed out physically, as *De ponderibus* indicates (in fact, here Jordanus describes how proposition 7 was discovered, namely, by subjecting proposition 2 to the *experimentum crucis*—does it work?).[31] The common algebra, introduced by the translations of Robert of Chester and Gerard of Cremona and expanded by Leonardo da Pisa, was adequate for the businessman at his counting tables and the lawyer with his inheritance problems.[32] But it was simply too pedestrian for the natural philosophers, despite its elegant expression in *De datis*. In short, the perspective of the thirteenth and fourteenth century natural philosophers apparently excluded algebra both as a subject for study and as a tool for problem-solving. They were not ready for analytic algebra.

Sources

A search for the well-springs of *De datis* is hardly rewarding. There are some indications of sources Jordanus might have tapped, but the signs are sparse. The authors whose works might yield source material are John of Seville, al-Khwārizmī, abū Kāmil, al-Karajī, Fibonacci, and Euclid.

Wherever Jordanus obtained his material, whether from his own fertile mind, from the production of others, or a combination of the two, he clearly shaped it according to his own plan. The format is stark: statement of the proposition, a general method for solving any equation that corresponds to the proposition, and an example demonstrating the application of the canonical form derived in the general method. Jordanus could have obtained this format from a reading of *Liber algorismi* of John of Seville (twelfth century).[33] Juan de Luna (as he is also called) devoted hardly more than a folio page of his book to a précis of algebra.[34] After indicating that he was excerpting passages from the book *gleba mutabilia* (sic), he stated three examples and developed each according to his plan. The three are: $x^2 + 10x = 39$, $x^2 + 9 = 6x$, and $3x + 4 = x^2$.

The book from which John purportedly quoted was *Kitāb al-jabr wa'l-mugābala* of al-Khwārizmī (died circa 850). "The Book of Restoration and Opposition of Number," as the author explained the title, set the pace for Arabic and European algebraists in succeeding centuries. It is a repository of Arabic algebra, showing briefly but clearly how to solve the three variations apiece of the two equations, $ax + b = 0$ and $ax^2 + bx + c = 0$. Geometric analogues rather than proofs make the exposé visual. Multiplication of polynomials, computing with radicals, and working with proportions complete the content brought into Europe by twelfth-century translators.[35] (This is important because, as the Rosen translation from the Arabic of the *Kitāb* indicates, there was a considerable amount of geometry in the original text of al-Khwārizmī.)[36]

It is difficult to believe that Jordanus was not familiar with some translation of al-Khwārizmī's *Kitāb* or his *Liber algebre*. Karpinski mentions two manuscripts in Paris, which were the basis for Libri's publication.[37] While these may have been in Paris during Jordanus' lifetime, there were enough translations (by Robert of Chester, Gerard of Cremona, and perhaps others) to warrant the assumption that Jordanus knew the book. It is also possible that he turned to *Liber algebre* for theorems; certain of its problems, in general form, become propositions in Jordanus:

al-Khwārizmī			Jordanus	
	(all assume $x + y = 10$)			(all assume $x + y = a$)
(p. 105)	$x/y = 4$	I-19		$x/y = b$
(p. 107)	$x^2 + y^2 = 58$	I-4		$x^2 + y^2 = b$
(p. 111)	$xy = 21$	I-3		$xy = b$
(p. 111)	$x^2 - y^2 = 40$	I-14		$x^2 - y^2 = b$
(p. 111)	$x^2 + y^2 + x - y = 40$	I-15		$x^2 + y^2 + x - y = b$
(p. 112)	$x^2 = 81y$	I-29		$x^2 = by$

Furthermore, *De datis* displays the three forms of the quadratic equation in the same order that they appear in *Liber algebre*:

"a square and roots equal to numbers" is
IV-8: $x^2 + bx = c$;

"a square and numbers equal to roots" is
IV-9: $x^2 + c = bx$;

"roots and numbers equal to squares" is
IV-10: $bx + c = x^2$

This arrangement seems hardly a coincidence. As in the alphabet, the sequence of equations may have been memorized. Jordanus had a place for them in *De datis*, and wrote them down as he recalled them. The proofs for the three propositions, as in *Liber algebre*, are established by the method known as completing the square. It is reasonable to assume, based on this evidence, that Jordanus was influenced by al-Khwārizmī's work.

Two other algebraic tracts may have served as sources for Jordanus: the *Kitāb fī al-jābr wa'l-muqābala* of abū Kāmil (circa 850–930) and the *Liber abaci* of Fibonacci (1202). Abū Kāmil was the algebraic successor of al-Khwārizmī; his commentary on the latter's text was in turn commented on, as early as the tenth century, by al-Istakhrī and al-'Imrānī.[38] His own work was translated into Latin, probably by Gerard of Cremona, toward the end of the twelfth century;[39] so it was available for Jordanus. The commonality of the *Kitāb* and *De datis* is disappointingly meager. A few problems of the first are similar to propositions in the other,[40] namely:

abū Kāmil		Jordanus	
(no. 5)	$x + y = 10$	I-3	$x + y = a$
	$xy = 21$		$xy = b$
(no. 7)	$x + y = 10$	I-14	$x + y = a$
	$x^2 - y^2 = 80$		$x^2 - y^2 = b$
(no. 3)	$x + y = 10$	I-19	$x + y = a$
	$x/y = 4$		$x/y = b$
(no. 4)	$x + y = 10$	I-29	$x + y = a$
	$x^2 = 9y$		$ay = x^2$

This is slight evidence to support abū Kāmil's work as a sourcebook[41] for *De datis*, particularly since the same problems appear in al-Khwārizmī.

Short shrift, it seems to me, should also be accorded Wertheim's suggestion that Jordanus adapted II-27 and II-28 from the algebra of al-Karajī (tenth and eleventh centuries), called *al-Fakhrī*. There seems to be no evidence that this work was translated into Latin, so Jordanus could not have drawn ideas from it.[42]

There is a close similarity between certain problems in the *Liber abaci* of Leonardo da Pisa (also known as Fibonacci) and *De datis*. For example:

	Fibonacci		Jordanus
(p. 410)	$x + y = 12$ $27y = x^2$	I-29	$x + y = a$ $ay = x^2$
(p. 411)	$x + y = 10$ $x^2 + y^2 = 62\frac{1}{2}$	I-4	$x + y = a$ $x^2 + y^2 = b$
(p. 416)	$x + y = 12$ $xy/(x - y) = 4\frac{1}{2}$	I-17	$x + y = a$ $xy/(x - y) = b$
(p. 454)	$x + y = 10$ $10/x + 10/y = 6\frac{1}{4}$	I-21.A	$x + y = a$ $c/a + c/y = b$

Levey, however, has pointed out that Fibonacci most probably borrowed these problems from the *Kitāb* of abu Kāmil.[43] I do not see this arguing for a Latin translation, as yet undiscovered and unrecorded, of the *Kitāb*, that may have been available to both Fibonacci and Jordanus. Leonardo could easily have learned of the problems during his travels in Arabian lands. At best, I would acknowledge a common source—namely, a compendium of problems similar to the various *Cautele* one finds in medieval mathematical codices.[44]

Several reasons prompt me to doubt any dependence of Jordanus on the *Liber abaci*. First, the two books are practically contemporaneous, and copies of the *Liber* were not broadcast over Europe: extent copies are apparently restricted to Italy. Second, Fibonacci consistently used one unknown (I have added the second unknown above). His first unknown is x, for example, and the second is $12 - x$. Jordanus clearly used several unknowns. Finally, if Jordanus had the *Liber* at his desk, he could easily have utilized far more examples than seem to appear in *De datis*. For instance, the transition is easy from this problem of Leonardo: "To one of two unequal quantities of which the one is thrice the other, I add its root. Similarly with the other quantity. And I multiply the two sums together and the result is ten times the larger quantity,"[45] to a pseudo-proposition for Jordanus: Given the ratio of two numbers and also the product of the sum of each with its root, the numbers can be found. This would have fitted appropriately in Book IV after proposition 7, where the ratio of two numbers is given together with the product of their sum and difference. In general, I doubt that Jordanus could have ignored the wealth of equations in chapter 15 of the *Liber*, had it been available to him.

I must acknowledge, however, that Jordanus might have taken from the *Liber* the idea to use letters to represent numbers. Recall that the *mos geometricus* used letters to represent line segments, which in turn represented numbers. Some of Fibonacci's letters skip the over the middle step, and refer immediately to numbers. For example, in discussing proportion he wrote, "Let *a*, *b*, *g*, *d* be four numbers in proportion, namely, *a* is to *b* as *g* is to *d*. Then, conversely, *b* is to *a* as *d* is to *g*."[46] Yet at the same time, Fibonacci's stated dependence on line segments to discuss number must be recognized. "The sciences of arithmetic and geometry are so connected, that each depends on the other. A person cannot discuss number without turning to geometry, nor geometry without using number."[47] Even if Jordanus took the idea from Fibonacci, however, he exploited it far more.

Regarding the use of Euclid as a source, one word leads me to suspect that Jordanus either knew Greek or had at hand some text translated directly from the Greek. The word is *latus*, which appears in IV-19, IV-20, IV-21, IV-22 and IV-23. Each of these propositions deals with squares and their roots. The Greek word for root is πλευρά, "side."[48] The Arabic translation for this was *jidr*; and it was commonly translated as *radix* (whence our "root") during the twelfth century.[49] The observation is important because Jordanus is not consistent in his use of either *radix* or *latus* where squares are concerned: in propositions both before and after those in Book IV, he uses *radix*. I have no firm answer for the questions, where did he get the term *latus*? and why did he use it in these five propositions?

Two works of Euclid, *Elements* and *Data*, would have been used as sources by Jordanus. A working knowledge of the *Elements* was a characteristic of every recognized natural philosopher;[50] the *mos geometricus* demanded this. And, as the *apparatus fontium* indicates, three propositions in *De datis* can be traced to the *Elements*, namely IV-7, IV-8, and IV-29. That the number is no greater, however, indicates that it was not a major source.

The more obvious source for Jordanus would seem to be the *Data*. Granted that it is concerned with given "areas, lines and angles to which others can be made equal,"[51] the similarity of titles prompts investigation. There is even a similarity in composition: fifteen definitions are followed by ninety-four propositions; in *De datis* three definitions are followed by one hundred and thirteen propositions. But a close look at the two basic things Jordanus might have borrowed, the method of proof and the individual theorems, makes clear that the *Data* was hardly a major source.

Jordanus' method of proof is not geometric, as is Euclid's. Where the *Data* displays the six steps of classic proof,[52] *De datis* presents an algebraic solution of a generalized problem. That is, an equation is subject to transformations until it has acquired a canonical form, which immediately supplies an indeterminate solution. Subsequently, a numerical example is shown which puts the canonical form to the critical test, as was discussed above in "Analysis and *De datis*."

Strong similarities exist between five propositions of *De datis* and a corresponding five in the *Data*. The English translation of II-2 and proposition 2 of the *Data* read the same: "If the ratio of a known number to some other number is given, then the other can be found." For practical purposes (and both books were intended to be practical), these pairs of of propositions are also equivalent: I-3 and proposition 85; I-5 and proposition 84; II-5 and proposition 5; II-7 and proposition 8; II-19 and proposition 7. These are possibly enough to warrant classifying the *Data* as a minor source for Jordanus. It may even have happened that a reading of Euclid's work suggested the idea of *De datis* to him, for both works are directed to the same goal: given certain information, to find what else is known.

In summary, then, no one work stands out as a major source that Jordanus might have used. Even assuming that all bits and pieces identified above were consciously taken from the respective works by Jordanus, he deserves to be regarded as an author, not a compiler, of a unique tract in advanced algebra.

Early Thirteenth-Century
Algebra and *De datis*

The search for sources provided a piecemeal conspectus of medieval algebra, and offers a perspective for evaluating Jordanus' contribution.

In western Europe during Jordanus' era, students of algebra learned from the books of al-Khwārizmī, abū Kāmil, and Fibonacci. (What John of Seville wrote could only whet their appetites, so his tract may be excluded from this discussion.) The available algebra consisted of simple and quadratic equations with one or several unknowns, together with methods for solving them; rules for operating with signed numbers, monomials, binomials, and roots; and finally parameters, false position, and the rule of three. After the basic material was developed with indenominate numbers, applications were customarily made to business and legal problems. Algebraic puzzles were also popular.[53] At least in England, efforts were made to generalize rules for solving numerical problems at an elementary level. The effect was to cast the word problem into the format of a simple equation for easy solution.[54] Some attempt was made to justify rules by appealing to geometric figures; but primarily the student was expected to memorize the rules.

The medieval student's focus of attention on solving equations differs diametrically from the focus today. Now we think in terms of the unknown; then they thought in terms of known numbers. For instance, in order to solve the equation, a square and 10 of its roots equal 39, abū Kāmil wrote, "Take half the (number of) roots; in this problem it is five. You multiply it by itself; it is twenty-five. Add this to thirty-nine; it is sixty-four. Take its root; it is eight. Subtract from it half the (number of) roots or five; three remains. It is the root of square."[55] (This is the same sort of emphasis, however, to which we expose students today who are learning to compute with calculators.) Any mystery concerning the origin or meaning of the foregoing verbal technique was immediately dispelled by abū Kāmil, who displayed the geometric method he learned from al-Khwārizmī. It is called, precisely, "completing the square." A square is drawn. Two rectangles of widths equal to the sides of the square and lengths equal to half the number of roots are affixed to adjacent sides of the square. The area of the resulting "carpenter's square" is 39, as given.

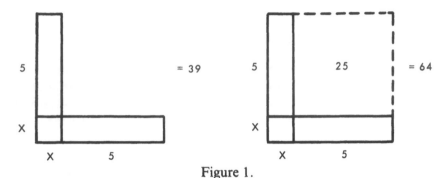

Figure 1.

The student is then instructed to fill out the L-shaped figure by adding a square 5 on a side. The addition "completes the square" and increases the area to 64. The *side* of the completed square is $x + 5$ and the *root* of 64 is 8. Hence, $x + 5 = 8$ and $x = 3$. (Negative roots, of course, were unacceptable, so many quadratics had only one root.) Note that the method is tantamount to using the formula

$$x = \frac{-(b) \pm \sqrt{b^2 - 4ac}}{2a}$$

which is of Babylonian origin.[56]

Let me emphasize here, because the remark is germane to the two italicized words above, that the term *root* is perhaps the most unfortunate misnomer in elementary mathematics. As explained in the section on sources, the original source of the word is the Greek πλευρά, which means "side." The Arabs translated this as *jidr*. By the time of the eleventh- and twelfth-century European translators, the Arabic word had assumed the additional meaning "root" of a plant or tree, which they rendered as *radix*. At the latest, during the Renaissance when the Greek manuscripts became available, the scholars translated πλευρά as *latus* or "side": this is the correct meaning of the word.[57] (Indeed, when the concept is first introduced in elementary school, it is pedagogically easier to define "root" as the side of a square.) Unfortunately, most people—including mathematicians—do not comprehend the denotation of the word "root."

Equally strange to us today is the descriptive terminology for the medieval problem-solving technique of false position. Essentially, this is a "guess and test" method. An entire book, *Liber augmenti et diminutionis*, exemplified the skill; it was probably composed in the eleventh century by someone named Abraham, whose identity is uncertain.[58] The *regula falsi,* as the technique is also called, consists of guessing the solution of an equation once or twice and then correcting for the error the "guess" produced; thereby the true value of the unknown is found. Abraham used the rule for nearly every problem in his book. It may be formulated thus: given any simple equation $ax + b = c$:

$$\frac{g_2e_1 + g_1e_2}{e_1 + e_2} = x,$$

where x is the unknown, g_n the guesses, and e_n the divergences from the true value of the equation. Jordanus applies the method in II-27 and II-28, where the problems consist of four equations with four unknowns, at a level of difficulty far exceeding anything in Abraham's text.

A final point must be considered. The Arabs and their medieval successors were ingenious at refining the concepts and skills they created or inherited, but all were hobbled by the *mos geometricus.* This is nowhere more evident than in the problem of finding the sum of two surds. Both abū Kāmil and Fibonacci correctly solve the problem,[59]

$$\sqrt{a + \sqrt{b}} + \sqrt{a - \sqrt{b}} = ?$$

They found the sum of the sides of two "irrational" or imperfect squares to be:

$$\sqrt{2a + 2\sqrt{a^2 - b}}.$$

The texts show that they were thinking in terms of geometric figures. Likewise, this mental framework forced them to envision a physical cube when they considered the sum of two cube roots. Fibonacci remarks, where he seeks the sum of the cube roots of five and three, ". . . relique vero radices, que proportionem non habent, (nec) addi nec disgregari possunt."[60] This mental pattern of visualizing problems as geometric forms prevented the solution of the cubic equation until the sixteenth century.[61] For, had either abū Kāmil or Fibonacci employed exactly the same *algebraic* techniques on the problem, to solve $\sqrt[3]{a + \sqrt{b}} + \sqrt[3]{a - \sqrt{b}} = x$, one of them would have been found the general solution of the cubic equation

$$x^3 = px + q, \quad \text{for} \quad p = 3\sqrt[3]{a^2 - b} \quad \text{and} \quad q = 2a.$$

Let us now consider the mathematics of *De datis*. Structural analysis of *De datis* reveals three introductory definitions and 113 propositions. The latter are separated into simultaneous and proportional equations and partitioned into four books. Each proposition has the same format: (1) a statement containing the relationship between what is given (*numeri dati*) and what is to be found; (2) an algebraic transformation of the relationship into a canonical form that supplies immediately the desired number(s); and (3) an example demonstrating the validity of the canonical form. Proposition II-19 will provide an illustration, in both verbal and symbolic translation. (I use x and y in the symbolic translation for the a and b of the literal translation.)

If the quotient of two parts of a given number is known, then the parts can be found. Let the given quotient of a and b be c. Increase this by 1 to get d. Because the product of b and c is a, then the product of b and d is ab. Therefore, divide ab by d to obtain b.

$$x + y = g$$
$$x/y = c$$

$$c + 1 = \frac{x}{y} + 1$$

$$= \frac{x + y}{y} = d$$

$$cy = x$$
$$dy = x + y = g$$

$$y = \frac{g}{c + 1}$$

And the canonical form receives immediate application:

For example, let the quotient of the two parts of 10 be 4 which is then increased by 1. Divide 10 by 5 to get 2, one of the parts.

$$y = \tfrac{10}{5}$$
$$y = 2$$

This example prompts two remarks. First, note in the English translation the symbol ab. This is, as Jordanus' text has it, addition by juxtaposition. The idea comes from addition of line segments by placing them next to one another. We still do it when we write a number in Arabic numerals: 53 is really $50 + 3$. Second, note the use of letters to represent numbers.

For the most part, Jordanus handles the letters well; there is almost no confusion. But two propositions, II-24 and II-25, tax his abilities: in these he uses, respectively, twelve and thirteen of the available letters. And in II-26 he exhausts all the available characters, leaving the reader with an impression of Babel.

The propositions in the four books fill the bulk of the text. Book I develops theories of simultaneous and quadratic equations, in twenty-nine propositions. It begins simply with the method for solving two simple simultaneous equations in two unknowns; from there it moves to four equations in four unknowns. Thereafter, and starting far more often with the given sum of two unknowns than with their difference, the propositions take the reader through many varieties of the second equation. They are, to mention a few, the given ratio of the two unknowns, their product, the sum or difference of their squares, the product of their sum and difference increased by the square of one of the unknowns, and the square of their difference. The canonical form is often either

$$x + y = m \quad \text{or} \quad x^2 + bx + c = 0$$

(in one of its varieties), and it is always understood that reference should be made to an earlier proposition to complete the solution. Occasionally, the canonical form yields one of the unknowns immediately.

Book II begins a theory of proportion in twenty-eight propositions, with the scope restricted to variations on four numbers in proportion. Here is a sampling: II-1, $a:x = b:c$; II-4, $x + y = a$, $(x + y):x = b$, $y:x = c$; II-10, $x + y = a$, $w + z = b$, $x:z = c$, $y:w = d$; II-22, $x + y = a$, $(x + c):y = b$. The canonical preference is for forms that yield either one of the unknowns, or the sum or difference of the two unknowns.

Each of the twenty-three propositions in Book III begins with at least three numbers in continued proportion; one or two of these, or the ratio of two, may be given. From proposition 5 on, they become highly complex. One or more numbers are given: either the sum of two or more unknown terms, or the ratio of the sum of two to the third, or the sum of the three and the ratio of two. There are five canonical forms: single unknown, ratio, product or sum of two unknowns, and quadratic equation.

Book IV picks up where Book II ended: four numbers, the last often being 1, in proportion but compounded, such as IV-2: $(x/y):(x/z) = a$, $y:z = b$. At the eighth proposition three quadratic equations appear, varieties of $x^2 + bx + c = 0$, either b or $c \neq 0$. These are followed by three proportional propositions reducible to a quadratic. Thereafter, save for the last, the propositions swing along a triangular path from proportions to simultaneous equations to quadratics, such as: IV-15, $(x + y):x = a$ and $(x^2 + y^2):z^2 = b$; IV-21, $x + y = a$ and $x^2y^2 = b$; IV-32,

$$(ax + c)(bx + d) = ex^2.$$

The text's single cubic equation concludes the book: $a:x^2 = b$ and $a^2:x = c$ become the cubic $x^3 = c/b^2$.

The foregoing textual analysis provides material from which an evaluation is fashioned. It appears that *De datis* was recognized as unique in its time, despite the *mos geometricus*. The number of copies, the spread of their geographic and temporal origins, and the codices in which they are housed, all suggest that both contemporaries and successors of Jordanus acknowledged the contribution *De datis* made to mathematics.

The number, spread, and origins of the copies are impressive. Fifteen copies are extant. Seven seem to be from the thirteenth century, six from the fourteenth and fifteenth centuries, and two from the sixteenth century. The geographic origins of these are, most probably, Italy (two), England (two),[62] Germany (three), and France (eight).

The codices containing the copies divide themselves into two groups. The first consists of quadrivial tracts, standard texts or references for students pursuing baccalaureate degrees. Two codices, the Ambrosiana (A) and the Ottoboniani (V), fill the first set. Both contain Jordanus' *Arithmetica* and *De datis*; A includes additionally Euclid's *Elements*. Since these MSS date from the fourteenth and fifteenth centuries, another conjecture suggests itself: university masters were becoming more demanding of their students. In addition to reading an algorithm and the first five books of Euclid, students were perhaps encouraged to read *De datis* also.

"Scientific anthologies" would best describe the second group, encompassing twelve of the codices. Some, such as Vienna 4770 (W), contain mostly mathematical tracts, including Robert of Chester's translation of al-Khwārizmī's *Liber algebre*.[63] Others, Basel F. II. 33 (B) for example, are collections of several hands and diverse dates. They include geographic, geometric, optical, astronomical, arithmetic, and algebraic works considered important in the late Middle Ages. In short, *De datis* keeps good company; however, as a reminder of the speculative nature of these remarks, I cannot omit noting that the final codex, BN 11885 (S), is an anthology that includes lives of the saints.

Had there not been an enduring respect for *De datis*, it seems inconceivable to me that this work could have received significant recognition by mathematicians two centuries after its composition. Digests and excerpts from *De datis* hold prominent places in late-fifteenth to mid-sixteenth-century academic endeavor: specifically, Johannes Widmann refers to it; Regiomontanus planned to publish it; Adam Riese quoted from it; and Johann Schuebel completely updated it. Such uses of the work suggest considerable appreciation for *De datis* in late medieval times.

Regiomontanus (1436–1476) died before realizing his plan to publish what he advertised as *Elementa arithmetica Jordani; data ejusdem arithmetica*. Four hundred years later, in 1879, Treutlein published a diplomatic transcription of the Basel MS which Curtze, three years later, criticized for some five hundred errors.[64] Still dissatisfied, the renowned paleographer of Thorn presented his own transcription of the Dresden DB 86 version in 1891.[65] This was translated into Russian by Schneider in 1959.[66] Apart from a few selections taken from other MSS to clarify issues raised by Curtze,[67] nothing of substance was offered until I prepared a critical edition for my doctoral dissertation in 1970.

Fonts for
the Critical Edition

In lieu of the autograph, as yet undiscovered, this edition makes use of the fifteen available MSS of *De datis*. Salient characteristics group them into two families, Alpha (α) and Beta (β), for stemmatic purposes. Four additional late MSS are described, witnesses to the popularity of *De datis* in the early Renaissance. Two final references from this period complete the inventory of the written tradition.

The fonts, of course, are the manuscript copies of *De datis*, valuable in and for themselves. The codices are listed alphabetically by sigla I found convenient. Where the information was available, I composed a table of contents for each codex. The reader will find author, title (in some cases, the incipit), and the Thorndike–Kibre incipit number for each work I could identify. All the works of Jordanus were matched against the scholarly catalogue of Thomson. Spellings of authors' names were those found, first, in *Dictionary of Scientific Biography*. Where this resource failed me,

I turned to *ISIS Cummulative Bibliography* for assistance. Under the rubric *special characteristics* I describe the utility of each MS for the construction of the critical edition. Dates and countries of origin of the codices are most often based on catalogue descriptions, but the dating and placing of the manuscripts of *De datis* were done by inspection of photographic copies.

A: Milano, Biblioteca Ambrosiana, Cod. lat. D 186 Inf. Italy, s. xiv in.
 1. Euclid, *Elementa.*
 2. Jordanus de Nemore, *De elementis arismetice artis.* TK 1600
 3. Jordanus de Nemore, *De numeris datis*, fol. 126vb–128vb. TK 959

The last item is written in two columns of forty-five lines each by a single scribe, in an Italian textual gothic hand of the first quarter of the fourteenth century. Plain initials and Roman numerals give little distinction to the text.

SPECIAL CHARACTERISTICS. The text is but a fragment of the work. While Book I is complete, Book II is broken abruptly in the middle of the demonstration of proposition 10, at the bottom of the verso of the last folio. The complement has not been found. Hence, this MS offered little assistance toward establishing the critical edition.

The codex belonged to Gian Vincenzo Pinelli (1535–1601) before it came to the Ambrosiana.

BIBLIOGRAPHY.

A. RIVOLTA. *Catalogo dei Codici Pinelliani (latini) dell' Ambrosiana* (Milan, 1933), 219, n. 296.

A. GABRIEL. *A Summary Catalogue of Microfilms of One Thousand Scientific Manuscripts in the Ambrosiana Library, Milan* (Notre Dame, 1968), 96, corrected by B. Hughes, "Toward an Explication of Ambrosiana MS D 186 Inf.," *Scriptorium* 1972, *26*: 125–127.

B: Basel, Oeffentliche Bibliothek, Cod. lat. F. II. 33, Germany, 1350–1380.
 1. Julius Solinus, *De situ orbis terrarum.* TK 1542
 2. Hyginus, *De ornatu coeli.* TK 661
 3. Anon., *De fluxu et refluxu maris.* TK 1703
 4. al-Farghānī, *Astronomia.* TK 960

5. Māshā'allāh, *Liber motus orbis et nature eius.* TK 722
6. Asculeus, *De ascensione signorum.* TK 1449
7. Jordanus de Nemore, *Arithmetica.* TK 1600
8. Gernardus, *Algorismus de minutiis.* TK 399
9. Nicole Oresme, *Algorismus de proportionibus.* TK 1596
10. Gernardus, *Algorismus de integris.* TK 431
11. ibn al-Haytham, *De speculis comburentibus.* TK 392
12. Pappus(?), *De figuris ysoperimetris.* TK 1083
13. Jordanus de Nemore, *Planispherium cum expositione anonymi.* TK 1119
14. Tideus, *De speculis.* TK 1388
15. ibn al-Haytham, *De crepusculis.* TK 1022
16. Thābit ibn Qurra, *Liber crastonis.* TK 260
17. Autolycus of Pitane, *De motu spere.* TK 1151
18. Banū Mūsā, *Liber trium fratrum de geometria.* TK 1687
19. al-Kindi, *De aspectibus.* TK 1013
20. Ps. Euclid, *De speculis.* TK 1084
21. John of Ligneres or Levi ben Gerson, *De triangulis.* TK 276
22. Anon., *Demonstrationes quadrantis.* TK 1305
23. Jordanus de Nemore, *De ratione ponderis.* TK 1000
24. Fredericus, *Quadratura circuli.* TK 1156
25. Theodosius of Bithynia, *De locis habitabilibus.* TK 684
26. Jordanus de Nemore, *De numeris datis*, fol. 138v–145v. TK 959
27. Jordanus de Nemore, *De triangulis.* TK 260
28. Johannes de Tinemue, *De curvis superficiebus.* TK 277
29. Dominicus de Clavasio, *Practica geometrie.* TK 1175
30. Theodosius of Bithynia, *De speris.* TK 1523
31. Levi ben Gerson, *De numeris armonicis.* TK 666
32. Campanus of Novara, *Theorica planetarum.* TK 1124
33. Thābit ibn Qurra, *De motu octave spere.* TK 661
34. Petrus de Mutina (Guclina?), *Theorica planetarum.* TK 1516
35. Ptolemy, *Perspective sive optica.* TK 287
36. Jābir ibn Hayyām, *Almagesti minoris libri vi.* TK 1006

Item 26 is written in fifty-three to fifty-six long lines each by a single scribe, in German textual and cursive gothic hands of the mid-fourteenth century. It has plain initials, roman and arabic numerals, and lacks marginalia.

SPECIAL CHARACTERISTICS. The scribe used two different scripts, one for the statements of the propositions, the other for demonstrations and examples. The theorems are numbered consecutively in fifteenth-century arabic numerals, with no separation into three or four books as in other MSS. But with one exception (I-24) all the numerals in the examples are roman. Unfortunately, the text abounds in errors and omissions, so much so that its contribution to an archetype is negligible. It was followed closely for the first two books, as the apparatus indicates; thereafter it was mostly ignored.

BIBLIOGRAPHY.

M. CURTZE, "Der Liber Trium Fratrum de Geometria," *Nova Acta d. ksl. Leop.-Carl. deutschen Akademie d. Naturforscher*, 1885, Ser. 2 *49*:111.

B. HUGHES, *Medieval Latin Mathematical Writings in the University Library, Basel* (private printing, 1971), n. 16.18.

C: Cambridge, University Library, Cod. lat. Gg. vi. 3. England, s. xiv.
1. Anon., *Variae tabulae astronomicae.*
2. John of Mauduith, *Tractatus super quatuor tabulis.* TK 1230
3. Anon., (incipit) "Quia ad noticiam celestis motus . . ." TK 1213
4. Roger of Hereford, *De iudiciis astronomie.* TK 1299
5. Zael Benbrit, *Introductorium ad astrologium.* TK 1411
6. Zael Benbrit, *De electionibus.* TK 985
7. Robert Grosseteste, *De spera.* TK 763
8. John of Lignères, *Equatorium plantearum.* TK 1106
9. Anon., *Varia astronomica.*
10. William Reade, (incipit?) "Hic quedam theorica utilis et brevis . . ."
11. Anon., (incipit) "Quedam notabilia experimenta theorice predicte(?) . . ."
12. Anon., *Variae tabulae.*
13. Campanus of Novara, *De quadratura circuli.* TK 136
14. ibn Tibbon, *Ars et operatio novi quadrantis* TK 143
15. Anon., *Tractatus de albion.*
16. Anon., *De novis numeris.* TK 1214
17. Anon., *Liber quantitatum mensurandarum per numerum.*

18. Anon., *De regulis generalibus algorismi ad solvendum omnes questiones propositas.* TK 1540

19. Anon., *Opus geometricum.* TK 794

20. Petrus de S. Audemaro, *Compositio instrumenti.* TK 1288

21. Anon., *Variae tabulae astronomicae.*

22. Jordanus de Nemore, *De numeris datis*, fol. 377v–381v. TK 959

23. Anon., *Variae tabulae astronomicae.*

Item 22 is written in thirty-nine to forty-three long lines each by a single scribe, in an English bookhand of the mid-fourteenth century. Plain initials and arabic numerals accompany the text.

SPECIAL CHARACTERISTICS. Books I and II are missing. Book III is complete and Book IV carries the reader through proposition 19. A footnote on the final folio states that the scribe knew where the remainder of the text lay. Nonetheless, the MS contains so many errors that it is of no assistance in preparing the critical edition.

The entire codex, once the property of the Benedictine Cathedral Priory of the Holy Trinity, Norwich, is a compilation of astronomical and mathematical works drawn from other codices.

BIBLIOGRAPHY.

A Catalogue of the Manuscripts Preserved in the Library . . . Cambridge (Cambridge, 1858), III, 215.[68]

N. R. KER, *Medieval Libraries of Great Britain*, 2d ed. (London, 1964), 136.

D: Dresden, Sächsische Landesbibliothek, Cod., lat. Db 86. France, s. xiii.

1. Euclid (trans. Adelard of Bath), *Elementa.* TK 586

2. Jordanus de Nemore, *De triangulis.* TK 260

3. Jordanus de Nemore, *De elementis arismetice artis.* TK 1600

4. Euclid, *Optica.* TK 957

5. Euclid, *Optica et catoptrica.* TK 7

6. Thedosius of Bithynia, *De speris.* TK 1523

7. Ps. Euclid, *De ponderoso et levi.* TK 266

8. Jordanus de Nemore, *Demonstratio de minutiis.* TK 1247

9. Proclus, *De motu.* TK 260

10. Jordanus de Nemore, *Demonstratio de algorismo.* TK 558

11. Archimedes, *Circuli dimensio*.   TK 996
12. Campanus of Novara, *De figura sectoris*.   TK 280
13. Cratilus, *Theoremata*.
14. Anon., *Theoremata geometrica*.
15. Anon., *De ysoperimetris*.   TK 1083
16. Euclid, *De ponderoso et levi* (not the same as item 7, above).
17. Anon., *De diametro*.
18. Anon., *De canonio*.
19. Anon., (incipit) "Si fuerit aliquod corpus . . ."   TK 1450
20. Jordanus de Nemore, *De ponderibus*.   TK 338
21. Johannes de Tinemue, *De curvis superficiebus*.   TK 277
22. Theodosius of Bithynia, *De habitationibus*.   TK 660
23. Campanus of Novara(?), *Demonstrationes pro astrolapsu*.   TK 1583
24. Anon., *De ortu signorum*.   TK 62
25. Euclid, *Liber de datis magnitudinibus*.   TK 363
26. Robert Grosseteste, *De quadratura circuli*.   TK 1156
27. Anon., *De circulo*.
28. Anon., *Secunda editio Tholomei*.   TK 1190
29. Anon., *De planisperio*.
30. Thābit ibn Qurra, *Introductio in almagestum*.   TK 502
31. Jordanus de Nemore, *Demonstratio de plana spera*.   TK 1524
32. Thābit ibn Qurra(?), *De proportionibus*.   TK 1139
33. Jordanus de Nemore, *De numeris datis*, fol. 228r–242v.   TK 959
34. Jordanus de Nemore, *De ratione ponderis*.   TK 1000
35. John Peckham, *Perspectiva*.   TK 769
36. Archimedes, *De incidentibus in humidum*.   TK 1295
37. Ps. Euclid, *De speculis* (excerpts).   TK 1084
38. ibn al-Haytham, *De speculis comburentibus*.   TK 392

Item 33 is written in forty long lines each by a single scribe, in a French textual gothic hand of the late thirteenth century. Ornate initials, roman and arabic numerals, and considerable marginalia characterize the text.

SPECIAL CHARACTERISTICS. The marginalia offer geometric representations of the numerical quantities in the demonstration, further evidence of the geometric bias of the times. The theorems are not numbered or separated into books.

The entire codex is the work of a single, expert scribe. The codex has similarities to Fournival's books. Once the property of a professor of mathematics, Valentine Thaus (circa 1570), it suffered considerable water damage during the air raids over Dresden in 1945.

BIBLIOGRAPHY.

M. CURTZE, "Ueber eine Handschrift der kgl. oeffent. Bibliothek zu Dresden," *Zeitschrift f. Math. u. Physik* (*Hist.-libt. Abteil.*), 1883, *28*: 1–13.

K. VON FALKENSTEIN, *Beschreibung der kgl. öffent. Bibliothek zu Dresden* (Dresden, 1839), 242.

F: Firenze, Biblioteca Nazionale, Conv. Soppr. J. V. 18 (Codex S. Marci Florentini 216). France, s. xiii ex.

   1. Anon., *Liber de umbris.* TK 1595
   2. Anon., (incipit) "Super illum locum Aristoteles in secundo de celo . . ." TK 1544
   3. Anon., *Liber ysoperimetrorum.* TK 1083
   4. [Possibly part of number 3], (incipit) "Isoperimetrorum ysoperimetrorum circulus comtemtorum quod est plurium . . ."
   5. Anon., *Vita secundi.* TK 1423
   6. Anon., (incipit) "Perisimetra sunt quorum latera coniunctim sunt . . ." TK 1035
   7. Anon., *Practica geometrie.* TK 870
   8. Jordanus de Nemore, *De triangulis.* TK 260
   9. Anon., *Liber de sinu demonstrato.* TK 477
   10. Anon., *Quadratura circuli per lunulas.* TK 1058
   11. Thābit ibn Qurra(?), *De proportionibus.* TK 1139
   12. Campanus of Novara, *De figura sectoris.* TK 280
   13. Jordanus de Nemore, *Demonstratio in algorismum* (*Communis et conseutus*). TK 958
   14. Jordanus de Nemore, *Tractatus minutiarum.* TK 875
   15. Jordanus de Nemore, *De numeris datis*, fol. 42v–53v. TK 959
   16. John of Seville, *Liber algorismi de practica arismetrice.* TK 1250
   17. Anon., (incipit) "Cum volueris almanac facere ad futurum annum arabum . . ." TK 352
   18. Robert Grosseteste, *De lineis angulis et figuris.* TK 1627
   19. Anon., *De numeris fractis.* TK 1475

20. al-Khwārizmī (trans. William de Lunis?), *Liber algebre.* TK 624
21. ibn al-Kamād, *De proportione et proportionalitate.* TK 1006

Item 15 is written in two columns of forty-seven to fifty lines each by a single scribe in a southern French or northern Italian textual gothic hand of the end of the thirteenth century. Well-formed initials, roman and arabic numerals, and extensive marginalia accompany the text.

SPECIAL CHARACTERISTICS. This is the earliest MS with the title as it is usually cited today: *Incipit liber Jordanis* (sic) *de numeris datis.* Proposition 23 of Book IV is completely missing. This was probably an oversight of the scribe, since its number is counted in the explicit. As detailed below, this MS is a copy of M.

Housed for some time in the library of the Dominicans of St. Mark, the codex has a number of Italianisms, such as *deo* for *deus* and *con* for *cum*. No scribes are identified.

BIBLIOGRAPHY.

A. BJÖRNBO, "Die mathematischen S. Marcohandschriften in Florenz," *Bibliotheca Mathematica,* 1880, ser. 3. *12*:220–221. (Nuova edizione a cura di Gian Carlo Garfagnini con una premessa di Eugenio Garin. Pisa, 1976 [Quaderni di Storia e Critica della Scienza, n.s.], 88–92.)

J: Krakow, Bibliojeteka Jagiellonska, Cod. lat. 1924. France, s. xiii.
1. Thābit ibn Qurra, *Introductio in Almagestum.* TK 502
2. Jābir ibn Hayyām, *Almagesti minoris libri vi.* TK 1006
3. Anon., *Secunda editio Tholomei.* TK 1190
4. Anon., (incipit) "Si a termina unius diametri circuli . . ."
5. Jordanus de Nemore(?), *Demonstrationes pro astrolapsu.* TK 1583
6. Theodosius of Bithynia, *De habitationibus.* TK 660
7. Thābit ibn Qurra, *De motu octave spere.* TK 661
8. Thābit ibn Qurra(?), *Introductio in Almagestum.* TK 502
9. Theodosius of Bithynia, *De speris.* TK 1523
10. Jordanus de Nemore, *Communis et consuetus.* TK 237
11. Jordanus de Nemore, *Tractatus minutiarum.* TK 875
12. Thābit ibn Qurra(?), *De proportionibus.* TK 1139
13. Jordanus de Nemore, *De numeris datis,* pp. 287–313. TK 959
14. Anon., "Tabula longitudinum".

Item 13 is written in forty-two to forty-five long lines each by a single scribe, in a French textual gothic hand of the mid-thirteenth century. The initials were not inserted. Roman and arabic numerals and a few marginalia accompany the text. Several pages are blank.

SPECIAL CHARACTERISTICS. Although the propositions are not numbered, the text is divided into four books by a space between the final proposition of one book and the first proposition of the next. The numerals in the text are roman; those in the marginalia are arabic. The marginalia are not geometric.

The codex itself, written in several hands, was for some time the property of Jacobo Piso (died 1527), secretary to the Cardinal Protector of Poland. The name of an earlier owner seems to have been scraped off page 3. Piso may have given it to Maciej Karpiga of Miechów (circa 1457–1523), longtime rector of the Academy of Krakow, from whose library the codex entered the university library. The codex has similarities to Fournival's books.

BIBLIOGRAPHY.

LESZEK HAJDUKIEWICZ, *Biolioteka Macieja z Miechowa* (Wroclaw, 1960), 406–408. The scholarly description of codex 1924 in this work improves greatly upon that found in Władysław Wisłocki, *Katalog Rekopisów Bibliojeteki Uniwersytetu Jagiellónskiego* ... (Krakow, 1877–81), 461.

M: Paris, Bibliothèque Mazarine, Cod. lat. 3642. France, s. xiii in medio.

1. Anon., *De natura terre et lapidum quorumdam.*
2. Thābit ibn Qurra, *Introductio in almagestum.* TK 502
3. Jordanus de Nemore, *Elementa super demonstrationem ponderum.* TK 1000
4. Petrus, *De compoto.*
5. Boethius, *Questiones de natura rerum.*
6. Herman the Lame, *De compositione astrolabii.* TK 611
7. William of Conches, *Philosophia.* TK 1307
8. Anon., *Varia astronomica.*
9. al-Khwārizmī (trans. Adelard of Bath), *Tabule astronomice* (Ezich). TK 822
10. Anon., *Arbor consanguinitatis.*
11. [Fragments] *De cognatione spirituali; De commendatione Aristotelis;*

12. *De Socrate.*
13. Jordanus de Nemore, *Demonstratio de algorismo* (*Communis et consuetus*).   TK 958
14. Jordanus de Nemore, *Tractatus minutiarum.*   TK 875
15. Jordanus de Nemore, *De numeris datis*, fol. 99ra–105ra.   TK 959
16. John of Seville, *Liber algorismi de practica arismetrice.*   TK 1250
17. Daniel, *Sompnia.*

Thereafter follow fragmentary extracts on matters philosophical and theological from such authors as Aristotle, Boethius, Cicero, and others.

Item 15 is written in two columns of fifty-five to fifty-seven lines each by a single scribe, in a French textual gothic hand of the mid-thirteenth century, probably in Paris. Plain initials, roman and arabic numerals, and extensive marginalia (Book I) characterize the text.

SPECIAL CHARACTERISTICS.   The marginalia go into great detail to explain and prove the various propositions. Furthermore, where the scribe erred or omitted something, the marginalia often provide the correct reading. This MS proved most useful for the critical edition. Probably once the property of a Parisian monastery, the codex is a compilation of several MSS of various dates and sources. My list of its contents presents a more detailed description than is found in the standard catalogue.

BIBLIOGRAPHY.
A. MOLINIER, *Catalogue des Manuscripts de la Bibliothèque Mazarine* (Paris, 1890), III, 152, n. 3642.

O: Oxford, Bodleian Library, MS Auct. F. 5. 28 (SC. 3623). England, s. xiii.
  1. Euclid (trans. Adelard of Bath), *Elementa.*   TK 1123
  2. Anon., (incipit) "Ex numeris sese respicientibus . . ."   TK 534
  3. Anon., *Liber de visu.*
  4. Euclid, *Liber de speculis.*   TK 1342
  5. Theodosius of Bithynia, *De speris.*   TK 1523
  6. Theodosius of Bithynia, *De locis habitabilibus.*   TK 660
  7. Campanus of Novara, *De figura sectionis.*   TK 1583
  8. Anon., *De ortu signorum.*   TK 62
  9. Euclid, *Liber de datis magnitudinibus.*   TK 363
  10. Dorotheus (Māshā'allāh?), *Liber de occultis.*   TK 600
  11. Jordanus de Nemore, *De numeris datis*, fol. 74v–87r.   TK 959

12. Theodosius of Bithynia, *De speris.* TK 1190

13. Thābit ibn Qurra, (incipit) "Equator circuli diei est circulus maior . . ." TK 501 (*sic*)

14. Jordanus de Nemore, *Elementa super demonstrationem ponderum.* TK 1000

15. Archimedes, *De quadratura circuli.*

16. ibn al-Kamād, *De similibus arcubus.* TK 583

17. Anon., *De ysoperimetris.* TK 1083

18. Anon., *Liber canonii.*

19. Anon., *Liber de canonico.* TK 1450

20. Johannes de Tinemue, *De curvis superficiebus.* TK 277

21. Robert Grosseteste (trans. unknown), *De quadratura circuli.* TK 1156

22. Gerard of Brussels, *De proportionalitate motuum et magnitudinum.* TK 1184

23. Jordanus de Nemore, *Elementa super demonstrationem ponderum.* TK 1000

24. Anon., *Geometria.* TK 149

Thereafter follow fifteen theological and philosophical treaties copied toward the end of the thirteenth century.

Item 11 is written in forty-three to forty-four long lines each by a single scribe, in an English (or French?) textual gothic hand of the third quarter of the thirteenth century. The initials were not completed. Roman numerals and some marginalia (Book I) accompany the text.

SPECIAL CHARACTERISTICS. While the propositions are numbered in a contemporary hand, the numbering differs from that found in other MSS. Books I and II are combined into a single set of fifty-six propositions; Books III and IV have the same numeration as the other MSS. As with M, the marginalia are confined to the first twenty-four propositions; unlike those in M, they are quite brief.

The codex represents a joining of two distinct sets of thirteenth century MSS, of English or French origin. On fol. 226v is a note stating that the codex was purchased in 1337 by Master John Cobulkdik of and for Oriel College.

BIBLIOGRAPHY.

R. MADAN et al., *Summary Catalogue of Western Manuscripts in the Bodleian Library at Oxford* (Oxford, 1937), II, pt. 2, n. 3623.

O. PÄCHT and J. ALEXANDER, *Illuminated Manuscripts in the Bodleian Library* (Oxford, 1973), III, n. 415.

N. R. Ker, *Medieval Libraries of Great Britain*, 2d ed. (London, 1964), 148.

P: Paris, Bibliothèque Nationale, Cod. lat. 8680 A. France, s. xiii.

1. Thābit ibn Qurra, *Liber de ponderibus.*   TK 260
2. Gerard of Brussels, *De proportionalitate et magnitudine motuum et magnitudinum.*   TK 1184
3. Jordanus de Nemore(?), *De ratione ponderis.*   TK 1000
4. Jordanus de Nemore, *De numeris datis*, fol. 11r–21r.   TK 959
5. Anon., (notes on the Ptolemaic system, in a slightly later hand).
6. Anon., *Liber de angulis.*
7. Gernardus, *Algorismus.*   TK 431
8. Campanus of Novara(?), *Demonstrationes pro astrolapsu.*   TK 1583
9. Anon., *De compositione quadrantis.*   TK 585
10. Pappus(?), *De figuris ysoperimetris.*   TK 1083
11. Ps. Euclid, *De ponderoso et levi.*   TK 266
12. Jordanus de Nemore, *Elementa super demonstrationem ponderum.*   TK 1000
13. Jordanus de Nemore, *Demonstratio de plana spera.*   TK 1524
14. ibn al-Haytham, *De speculis comburentibus.*   TK 392
15. Apollonius of Perga, *De pyramidibus* (from: *De conicis*)   TK 287
16. Anon., (In a much later hand three folio sides containing a commentary on the *Book of Sentences*).

Item 4 is written in forty-five to forty-six long lines each by a single scribe in a French textual gothic hand of the third quarter of the thirteenth century. Ornate initials, roman numerals, and extensive marginalia (perhaps from the early fourteenth century) accompany the text.

SPECIAL CHARACTERISTICS.   A single example of a hindu-arabic numeral is found on fol. 15, line 10, a thirteenth-century 4 (ℓ). I suspect that this was a scribal lapse, since all other numerals are roman. A blank folio, 22v, follows the text. Because of this, I would hazard that the scribe intended to complete the text of *De datis* but never got around to it. The text is good, and quite useful for the critical edition, though difficult to read from the microfilm.

The codex was once in the library of the mathematician Claude Hardy (died 1678), from whence it passed to the collection of Jean Colbert (1619–1683). On folio 65v is a note, in a fourteenth-century hand, that it was purchased for "IIII escus d'or." Except for item 5, which is in a later hand, the codex is a complete unit penned by one scribe, with florid initials introducing the various tracts.

BIBLIOGRAPHY.

Catalogus codicum manuscriptorum Bibliothecae Regiae (Paris, 1739–44), IV, 534, c. 2—appendix.

For the library of Claude Hardy, see Paul Colomiès, *Opera* (Hamburg, 1709), 165–166 and 259–260.

R: Paris, Bibliothèque Nationale, Cod. lat. 11863. France, s. xvi.
  1. Jordanus de Nemore, *De numeris datis*, fol. 1r–18r. TK 959
  2. Alexander Anderson, *Appendix pro calculu motuum quinque planetarum.*
  3. *Tractatus de commensurabilibus et incommensurabilibus.*
  4. Maur Fouguet, *Reflexions sur la censure de la gnominique par le calcul et par la géométrie.*
  5. Marinus, *Hypomnema.*

Item 1 is written in two columns of thirty-three to thirty-seven lines each by a single scribe in a French cursive hand of the mid-sixteenth century. Plain initials and roman and arabic numerals accompany a text that lacks marginalia.

SPECIAL CHARACTERISTICS. The tract has a double title: *Data numerorum Jordani*; *seu Jordani de datis numeris*. This is noteworthy because the hand of the second title differs from that of the first, which is the hand of the text. The principal scribe had a neat hand, almost devoid of abbreviations. Since this MS seems to be a copy of P, it does not enter into the critical edition.

Both items 1 and 3 are "ex bibliotheca Lustierina" and are now part of the Saint-Germain collection.

BIBLIOGRAPHY.

L. DELISLE, *Bibliothèque de l'École des Chartes* (Paris, 1865), 6 série, I, 209.

L. DELISLE, *Inventaire des manuscrits de Saint-Germain-Près* (Paris, 1868), n. 11863.

S: Paris, Bibliothèque Nationale, Cod. lat. 11885. France, s. xiii.
  1. Jordanus de Nemore, *De elementis arismetice artis.* TK 1600
  2. Jordanus de Nemore, *De numeris datis*, fols. 110^{bis}b–112b (*olim* 24b–26b; note, fol. 110bis is misnumbered as fol. 100bis), 115 (*olim* 27; also numbered as fol. 114). TK 959

Our MS is written in two columns of fifty-three lines each by a single scribe, in a southern French gothic hand of the mid-thirteenth century. It appears to be a university manuscript. Common initials, roman numerals, and a few marginalia in contemporary and later hands accompany the text.

SPECIAL CHARACTERISTICS. The text is incomplete; it ends toward the conclusion of II-10. The appearance of the final folio gives the impression that the scribe was interrupted at his work and never returned to complete it. The marginalia contain both roman and arabic numerals without geometric significance. A much later hand identifies the tract, "Data numerorum eiusdem Jordani."

The codex is a collection of hagiography in Latin (s. xiii) and tracts on French history in French (ss. xv and xvi), with Jordanus' works sandwiched between. Both Jordanus' works were penned by the same scribe. In a fourteenth-century hand on fol. 111 is the note, "Viro venerabili et discreto magistro Petro Archerii scolastico Ariensi vices gerens reverendi in Christo Patris domino Silvanectensis episcopi conservatoris pri(vilegiorum)."

BIBLIOGRAPHY.

L. DELISLE, *Inventaire de manuscrits de Saint-Germain-Près* (Paris, 1868), n. 11885.

T: Vatican, Bibliotheca Apostolica, Cod. Vat. lat. 4275. France, ss. xiv and xv.
  1. John Cusinus, *De sufficientia legis Christiane.*
  2. Anon., (incipit) "Casus sequentes tangunt . . ." (perhaps a commentary on the preceding).
  3. Albert the Great, *Speculum astronomie.* TK 975
  4. Robert de Bardis, (incipit) "Quesitum fuit utrum per interrogationes astronomicas . . ." TK 1199
  5. Nicole Oresme, *Contra astrologos.* TK 887
  6. Anon., (incipit) "Utrum stelle videantur ubi sunt . . ."
  7. Anon., (gloss of the preceding).

8. Jordanus de Nemore, *Demonstratio de algorismo.* TK 558

9. Jordanus de Nemore, *De numeris datis,* fol. 70ra–81vb. TK 959

10. Archimedes, *Circuli dimensio.* TK 996

11. Thābit ibn Qurra, *De motu octave spere.* TK 661

12. Nicole Oresme, *Algorismus proportionum.* TK 1596

13. Anon., (incipit) "Cum de tribus propositis, duo utrumque . . ."

14. Nicole Oresme, *De proportionibus proportionum.* TK 1002

Item 9 is written in two columns of forty to forty-one lines each by a single scribe, in a southern French hand of the third quarter of the fourteenth century. Plain initials and arabic numerals typify the text, which lacks marginalia.

SPECIAL CHARACTERISTICS. This MS together with J and S came to my attention several years after the others had been collated. It adds nothing to previous work, apart from further testimony to the interest in *De datis.* The text is garbled with incorrect transcriptions and omissions. Finally, the script and disposition of the pages are identical with a fourteenth-century papal register from Avignon (see Vat. lat. 12111).

Of unknown provenance, the entire codex is a compilation from three sources: items 1 to 7 are in long lines, items 8 to 11 in two columns, and items 12 to 14 in long lines again. The last set is the oldest, the middle group is next, and the first set is the youngest. Three hands, one for each group, signal three distinct scribes.

BIBLIOGRAPHY.

Sala di Consultazione dei Manoscritti, cccv, n. 4275 (Vatican Library unpublished inventory, n.d.)

V: Vatican, Biblioteca Apostolica, Fondo Ottoboniani, Cod. lat. 2120. Italy, ss. xiv ex. to xv in.

1. Jordanus de Nemore, *De elementis arismetice artis.* TK 1600

2. Jordanus de Nemore, *De numeris datis,* fol. 96v–124v. TK 959

The second item is written in twenty-seven long lines each by a single scribe, in an Italian hand of the late fourteenth or early fifteenth century. It lacks initials. Arabic numerals and some marginalia accompany the text.

SPECIAL CHARACTERISTICS. Marginal glosses divide the tract into four books and not three, as the explicit states. Despite the fact that the text has many errors and omissions, it does show all the propositions, so it was consulted closely while editing Book IV.

The codex came from the library of Giovanni Angelo, Duke of Altemps (died 1627).

BIBLIOGRAPHY.

J. DALE and C. ERMATINGER, "Mathematics in the Codices Otto-boniani Latini," *Manuscripta* 1964, 9:21.

W: Wien, Nationalbibliothek, Cod. lat. 4770. Germany, s. xv.
  1. al-Khwārizmī, *Liber restaurationis et oppositionis numeri.*
     TK 455
  2. Jordanus de Nemore, *De numeris datis*, fol. 13r–40r. TK 959
  3. Euclid(?), *De elementis* (incomplete). TK 1152
  4. Anon., *Carmen quadripartitum de mathematica cum commentario.*
     TK 106
  5. Anon., (incipit) "Marcha est limitata ponderis . . ." TK 848
  6. Anon., *Notabilia de algorismo proportionum.*

Item 2 is written in thirty-one to thirty-two long lines each by a single scribe, in a German cursive hand of the mid-fifteenth century. Plain initials and arabic numerals mark the text, which lacks marginalia.

SPECIAL CHARACTERISTICS.   The MS contains five items not found in earlier copies. The first, between 1-16 and 1-17, is a statement of the commutative law for multiplication. The other four items are unique proofs for propositions 1-3, 1-4, 1-5 and 1-23. As the apparatus and discussion below indicate, W makes its own lineage within the Beta family of MSS. Unfortunately, it is replete with errors and omissions, so that it was useless for the critical edition.

Once the property of an undetermined "Magister Henricus Leo," the codex entered the Nationalbibliothek in 1780.

BIBLIOGRAPHY.

Tabulae codicum . . . in Bibl. Palat Vindobon. asservatorum (Wien, 1870), III, n. 4770.

X: Wien, Nationalbibliothek, Cod. lat. 5303. Germany, ss. xv and xvi
  1. Gerard of Brussels, *De proportionalitate motuum et magnitudinum.*
     TK 1184
  2. Johannes de Tinemue, *De curvis superficiebus.* TK 277
  3. Anon., *Propositiones tredecim de doctrina sinuum.* TK 231
  4. Anon., *Modus specula praeparandi.*

5. Euclid, *Optica et catoptica*. TK 7
6. Blasius of Parma, *Demonstrationes geometrie in theorica plane-tarum*. TK 200
7. Jordanus de Nemore, *De numeris datis*, fol. 87r–98r. TK 959
8. Anon., *De numero stadiorum uni gradui circuli majoris cor-respondentium varia notabilia*.
9. George Peurbach, *Fabrica et usus instrumenti pro veris coniunc-tionibus et oppositionibus solis et lunae*. TK 282
10. Anon., (incipit) "Instrumentum ad lineam meridianam..." TK 753
11. ibn al-Haytham, *De crepusculis*. TK 1022
12. Anon., *Commentarii in theoricas planetarum a Georgio Peurbach editas*. TK 391
13. ibn Tibbon, *De compositione novi quadrantis et de eiusdem utili-tatibus*. 827
14. G. Marchio, *Tractatus quadrantis planitorbii*. TK 828
15. Anon., *Tabula declinacionum solis secundum Ptolomaeum et secundum Alfaragium*.
16. George Newburgh, *Tabulae elevationum solis ad Viennam, Noribergam, Pragam, Venetias et Newburgum accomodatae*.
17. John of Gmuden, *De compositione cylindri*. TK 1580
18. Anon., (incipit) "Horologium concavum et in dimidia..." TK 640
19. Anon., (incipit) "Habito vase interius rotundato..." TK 595
20. George Newburgh, *Tabula linearum pro horis inequalibus*.
21. Anon., *De compositione chilindri*. TK 776
22. Anon., (incipit) "Confectio horologii contra murum.." TK 245
23. Anon., *Modus construendi horologium in muris*. TK 1020
24. Anon., (incipit) "Torcular sic construitur..." TK 1575
25. John of Lignères(?), *Modus conficiendi instrumentum pro linea meridiana invenienda*. TK 753
26. Anon., (incipit) "Instrumentum ad capiendum..." TK 752
27. Anon., (incipit) "Ad faciendum horologium horarum.." TK 39
28. Johannes Schindel, *Compositio chilindri*. TK 203
29. Anon., (incipit) "Horologium Achab collocare..." TK 640

30. Anon., *Collectanea astronomica.*
31. Anon., (incipit) "Accipe tabulam planam . . ."   TK 25—(?)
32. Anon., *Modus construendi horologium in muris ad meridiem.*
33. Anon., *Modus conficiendi horologium horizontale.*
34. Richard of Wallingford, *Albion.*   TK 74

Item 7 is written in forty-four to fifty long lines each by a single scribe, in a German cursive hand of the first third of the sixteenth century. Plain intials together with roman and arabic numerals accompany the text. There are no marginalia.

SPECIAL CHARACTERISTICS.   This is quite obviously a copy of W, both in correct and incorrect readings. It does lack III-3, which is in W. The propositions are not numbered or divided into books. The MS was of no value toward the construction of the critical edition.

The codex came from the Fugger Library to the Hofbibliothek in 1656.

BIBLIOGRAPHY.

Tabulae codicum . . . in Bibl. Palat. Vindobon. asservatorum (Wien, 1870), IV, n. 5303.

Two Families of Manuscripts

As the *special characteristics* suggest, obvious internal evidence fixed criteria for separating the MSS into two families, Alpha and Beta. The major criteria are

M-1: the MS contains either 95 or 113 propositions (abbreviated Ths in the following matrix).

M-2: the MS contains either propositions I-21α, I-22α, I-28α, II-12α, II-13α, and II-14α or a corresponding β set of propositions (both sets appear in the critical edition).

M-3: IV-15 either is complete or terminates at line 9.
The minor criteria are

m-1: III-23 has either the word *diminuendo* or the word *continuendo* (line 15).

m-2: IV-10 has either the word *dato* or the word *quadrato* (line 2).

m-3: IV-15 has either the word *similiter* or the word *simul* (line 2).
The following matrix collates MSS with criteria. A clear pattern the separation of the manuscripts into their respective families.

Criteria	MSS	13th c.							14th c.			15th c.				16th c.
		M	O	D	F	J	S	P	A	B	C	T	V	W	X	R
M-1	95 Ths		x	x			1	x	2	x	3	x				x
	113 Ths	x			x	x							x	x	x	
M-2	Set α		x	x				x		x	4	x	x			x
	Set β	x			x	x	x		x					x	x	
M-3	incomplete		x	x	x	x	x	x		x	x	x	x	x	x	x
	complete	x														
m-1	diminu'do	x	x	x	x			x		x		x	x			x
	contin'do	x	x	x	x	x					x		x		x	x
m-2	dato	x	x	x	x	x		x		x			x			x
	quadrato	x									x	5				
m-3	similiter		x	x	x	x		x		x	x	x	x		x	x
	simul	x		x										x		x
	Family	β	α	α	β	β	β	β	β	α	β	α	α	β	β	α

1. MS stops at end of II-10.
2. MS stops within II-10.
3. MS contains only Book III (complete) and Book IV to proposition 19. Since the scribe wrote that he knew of the rest of the text, this MS is classified as complete.
4. Criterion is inapplicable because it refers to theorems in Books I and II which are missing in this MS.
5. Neither word appears in the MS.

While perfect symmetry in the matrix would have made the family memberships quite obvious, the lack of perfection finds a probable explanation in the criteria themselves. Regarding m-3: the abbreviations for *similiter* and *simul* are the same except that the former has an elevated "comma" to the right of the *L* and the latter has it to the left; a slip of the pen changes the denotation of the abbreviation. Supposing this to have happened, one may assume that J, C, D, F, and X fit their respective familial patterns; R has *simul* written out.

Reaching a familial assignment for V presented its own difficulties: the MS has criterion M-3 characteristic of the Beta-family but all minor criteria of Alpha. However, M-3 is implicit in M-1 because incompleteness means that IV-15 breaks off in the middle of the sentence (line 9). No MS in Alpha concludes with the complete IV-15. Thus, the difference between V and the Alpha set is that V has all the theorems. I suspect that V testifies to a less contaminated strain in the Alpha family and have classified it accordingly.

Intrafamilial Relationships

The chronological spread of the manuscripts and their respective variants suggest intrafamilial relationships from which I constructed the stemmatization.

In the Alpha family, D, O, and P are from the thirteenth century, B from the fourteenth, T and V from the late fourteenth or early fifteenth, and R from the sixteenth. An inspection of P and O indicates that they have a number of errors in common (e.g., in I-3, I-24, I-28α). On the other hand, each has its own set of errors. P clearly varies[69] in I-2, I-13, II-16, and II-20 from O; and O varies from P in I-5, I-17, I-20, and II-3. Hence, while P and O have a common source, neither was the source of the other. The incomplete T, however, does have a number of readings unique to O and lacks those peculiar to B, D, and P. The scribe of T was not copying from V because after stopping abruptly in the middle of IV-15 he continued with Archimedes' *De mensura circuli* (incipit: "Omni circulus triangulo orthagonus est equalis"). Hence, I put T in line with O.

A comparison of B with O and P shows correct readings in I-24 and I-28 where the two last err; but B has its own error in II-11. B has none of P's unique variants and each corresponding reading is correct in B. Hence, there seems no connection between B and P. Despite shortcomings,

B apparently has none of O's errors. Either the scribe was able to correct the variants as they were discovered in O or he had an exemplar different from O. Furthermore, B supplies some major omissions of O (e.g. I-17 and I-25). Hence, I judge that B was copied independently of either O or P. Whatever errors it has in common with them are from their common archetype.

The condition of D, particularly on microfilm, precludes any close study. However, quite evidently there is an additional theorem in Book I. It is between theorems 16 and 17 and states, "Si enim fuerint duo numeri inaequales, quadratum maioris maius est multiplice eiusdem a minori denominatione, quantum est, quod fit ex eorundem differentia in se ipsum et in minorem ducta."[70] Moreover, I-26 has a lengthy insertion calling the reader's attention to a "beautiful corollary." Both of these are absent from B, O, and P. It appears that D is a fourth line in the Alpha family.

Apart from considerations appearing on the matrix, there are several reasons for listing R as a distant descendant of P. First, R does not have the insertion found in D between I-16 and I-17; this seems to eliminate D as a predecessor of R. Also, R varies from B and O exactly as P; consider, for instance, I-2, I-15, II-11 and II-17. Finally, the differences between P and R seem minor.

The final member of the Alpha family seeks a separate classification, as suggested above. V is the only member that has all the propositions. While it has some of the Alpha characteristics and certain variants in common with the other members of the family as the table indicates, I conclude that V entered the family collaterally.

Hence, the stemma proposed for the Alpha family is this:

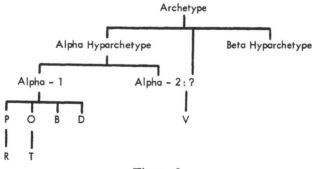

Figure 2.

The stemmatization of the Beta family can be treated with more dispatch. Chronologically, M appears to be the oldest manuscript and it is complete. C is incomplete at both beginning and end. Assuming that the scribe of C was not interested in the first two books, he finished IV-19 and stopped. His exemplar was apparently incomplete, as he remarked in a footnote that the remainder of the tract could be found "elsewhere."[71] This would be impossible were he copying directly from M. Despite the presence of a number of variants in common with M, testifying to a common ancestor, C was not taken from a linear decendant of M. This may be concluded from an inspection of III-1, IV-14, and IV-15: in III-1 C has *Quod sic probatur*, which is lacking in M; in IV-14 C has *ex illis*, which M omits in the statement of the proposition; and in IV-15 C has *similiter* where M has *simul*. Therefore, C is collateral with M.

In both M and F, Jordanus' *De algorismo* precedes *De datis*. Moreover, the marginal and interlinear glosses in the two MSS correspond word for word, except that there are more in M. IV-23 is missing in F, but this is not noticed in the explicit, which totals the propositions as though it were present. Finally, F lacks a sentence in IV-29 that is present in M. The conclusion is clear, that M is a direct linear ancestor of F.

The Viennese MSS W and X are members of the Beta family, but not in the same way as M, C, and F. Although the Viennese MSS have some of the characteristics of these three, they do not have them all. In particular, the two have a theorem inserted between I-16 and I-17; they both lack II-5; and they have their own common proofs for I-3, I-4, and I-5, and a different I-23. These variants argue for a partition of the Beta family. Supportive evidence for this partition is found in the digests discussed below, Dresden C 80 and Wien 5277, which have *both* versions of I-23! W and X do not. Thus there must be a Beta-1 and Beta-2, with W and X assigned to Beta-2.

Manuscripts A, J, and S remain. All contain the I-23* found only in Beta-2 but not the proofs for I-3, I-4, and I-5 unique to W and X. Rather, these proofs in A and S agree substantially with those in J, which alone of the three contains all the theorems. A and S break off in II-10, but at different places. Nonetheless, both A and S are from the same source, since they contain the same incorrect reading that begins the proof of I-24. Since J has the correct reading (as well as all the theorems), I believe its origin to be distinct from the others in the Beta-2 strain.

Consequently, the Beta family can be partitioned as follows:

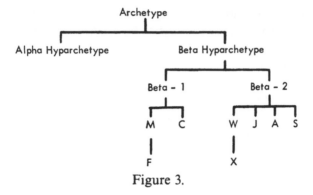

Figure 3.

The composite stemmatization of all known MSS is this:

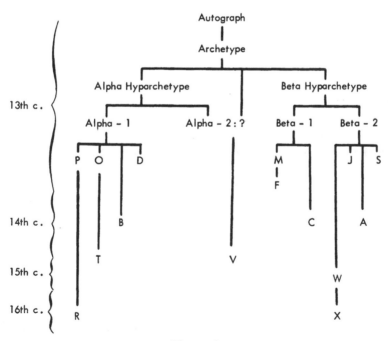

Figure 4.

Digests and Excerpts

During the transition from the Middle Ages to the Renaissance, four manuscripts from the genre of digests witnessed a different tradition of *De numeris datis.*

1. Dresden, Sächsische Landesbibliothek, Cod. Lat. C 80, fol. 316r–323r (dated: 1484).

DESCRIPTION. Franz Schnoor von Carolsfeld, *Katalog der Handschriften der Königl. öffent. Bibliothek zu Dresden* (Leipzig, 1882), I, 198.

SIGNIFICANCE. This was the MS used by Curtze to complete Book IV of his transcription of *De numeris datis.*

2. Göttingen, Universitätbibliothek, Cod. Lat. Philos. 30, fol. 151v–184v (1545–48).

DESCRIPTION. *Verzeichniss der Handschriften im Preussischen Staate* (Berlin, 1893), I, 144.

SIGNIFICANCE. Three pages of introductory material discuss subtraction and the types of quadratic equations that can be built upon subtraction, multiplication of binomials that result in quadratic equations, and division. The author found it necessary to remark that theorems 21 and 13 (in that order!) are useful for solving other problems.

3. Wien, Nationalbibliothek, Cod. Lit. Nr. 5277 (Philos. 68), fol. 320v–327r (s. xvi).

DESCRIPTION. *Tabulae codicum . . . in Bibl. Palat. Vindobon. asservatorum* (Wien, 1870), IV, n. 5277.

SIGNIFICANCE. After a notable explicit (*Explicit de datis non datum gratis*), another writer, the one who glossed much of the text, began but did not complete an analysis of the preceding.

4. New York, Columbia Universty Library, X 512, Sch. 2.q. (313450), pp. 160–309 (1550–55).

DESCRIPTION. Barnabas B. Hughes, O.F.M., "Johann Scheubel's Revision of Jordanus de Nemore's *De numeris datis*: An Analysis of an Unpublished Manuscript," *Isis* 1972, *63*:222–24.

SIGNIFICANCE. Except for the statements of the propositions and some of the examples, this is an original work by Scheubel, a complete revision of Jordanus' tract.

Internal evidence suggests a hypothesis about the purpose of these digests and a direction for stemmatization. Scheubel was clearly writing a text; his revision is but part of a projected textbook on algebra. The Wien and Göttingen MSS may well have been lecture notes for an instructor in the "great art." Their introductory material and analyses evince scholarly reflection. The bareness of the Dresden MS hints at its being nothing but an exercise. All these uses show that scholars were taking *De datis* seriously.

Three shared items point to one exemplar for these digests: the complete number of propositions, the Beta set of propositions, and the theorem introduced between I-16 and I-17 and equivalent to the commutative law for multiplication. Only W and its offspring X have these characteristics, so one of these must be the exemplar. Peculiar rearrangements of the propositions within the respective digests suggest the split shown in this stemmatization:

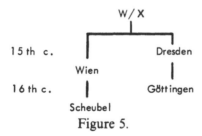

Figure 5.

Not to be overlooked are two final witnesses to the esteem in which *De numeris datis* was held by scholars of the late fifteenth and early sixteenth centuries: Johannes Widmann (1460–?) and Adam Riese (1489?–1559). The former is credited with introducing algebra, as a distinct course, for the first time to a university audience, in his lectures at Leipzig in 1486. He referred to *De datis* by name and quoted the statement, but not the proof, of II-10.[72] Riese, Germany's best known *Rechenmeister*, wrote a *Coss* (algebra text) in which he included German translations of some fourteen propositions from Books I and II.[73]

Methodology

The purpose of the critical edition is to discover the algebra written by Jordanus de Nemore. Although neither the autograph nor the archetypal MS has been found, initial collating of all fifteen MSS demonstrated the superior reliability of three, P, O, and M, with significant assistance from J and V for Book IV. The slight use of the other ten MSS is justified under their respective "Special characteristics" sections.

Several editorial liberties have been taken. Regardless of original numeration or lack thereof, each proposition in the critical edition has been assigned a roman numeral to indicate its book and an arabic numeral to identify its order within the book. Propositions unique to families are identified by the letters α and β. Lacking the autograph, which would indicate Jordanus' orthographic preferences, spelling has been standardized. The representation of numbers presented a difficulty, but I resolved it in this manner: since the MSS cite numbers both by words and by roman and arabic numerals, either the name or the numeral for whole numbers was adopted as the MSS indicated, slight variations being ignored in the apparatus. As for fractions, the *name* was arbitrarily made a word, its *number* a roman numeral. This decision relieved me of having to cope with such varieties as "iii iv," "iii quarte," and "tres quarte," not to mention the "$\frac{3}{4}$" of later MSS, all of which appear below as "iii quarte."

Finally, I should remark on punctuation. A number of manuscripts distinguish between the statement of the proposition and its demonstration and example by a different-sized script. For clarity I capitalized all statements. For similar reasons, each proposition is one paragraph. The medieval practice of placing a dot before and after numerals, to prevent their change, is not reproduced. Letters used for numbers are italicized. There was no harmony among the MSS with respect to initial capitalization and punctuation, so I adopted those forms that contributed most to the mathematical sense of the text. All these changes reflect my belief that the mathematics of Jordanus de Nemore concerns his readers, not the grammatical expertise of the copyists.

The Symbolic Translation

Following the critical edition is a translation of the text into modern algebraic symbols. The format of the symbolization together with several liberties taken by me require some explanation.

Four columns set off the general format. The first identifies the book and theorem number. The second couches the hypothesis in the appropriate number of equations. The third shows the reduction of the hypothesis to the canonical form reached by Jordanus, which exposes either the unknown(s) or the penultimate step. Where the last is the canonical form, the fourth column refers the reader to the pertinent proposition(s) whereby the unknown(s) is ultimately found. Note, however, that this final reference is made vaguely by Jordanus; e.g. "... ex premissis" in I-11. Hence, the fourth column is my offering for completeness.

The liberties are predictable. First is the Cartesian convention: letters from the beginning of the alphabet represent constants, and from the end signify unknowns. Jordanus did not distinguish between given and unknown quantities where using letters; the context clarifies their meaning. Second, some of the canonical forms I construct go beyond what Jordanus stated. For instance, IV-14 in the text stops at $(x - y):z$. This is unclear, so I show that the actual unknowns are $x:z$ and $y:z$.

Notes to Introduction

1. B. Thomson, "Jordanus de Nemore: Opera," *Mediaeval Studies*, 1976, *38*:97–144 completely supersedes all previous discussions on the identification and ascription of the works of Jordanus.

2. See for instance E. Grant, "Jordanus de Nemore," in *Dictionary of Scientific Biography* (New York: Charles Scribner's Sons, 1973), VII, 171–72; B. B. Hughes, "Bibliographic Information on Jordanus de Nemore To Date," *Janus* 1975, *62*:151–56; A. G. Molland, "Ancestors of Physics," *History of Science* 1975, *12*:64–67; R. B. Thomson, *Jordanus de Nemore and the Mathematics of Astrolabes: De plana spera* (Toronto: Pontifical Institute of Medieval Studies, 1978), 1–17.

3. Chancellor of the Cathedral in Amiens, Richard de Fournival composed a list of works desirable for the library he planned for scholars— a medieval "great books" program. A facsimile of this MS has been published by J. J. de Vleeschauwer, *La Biblionomia de Richard de Fournival du Manuscrit 636 de la Bibliothèque de la Sorbonne. Texte en facsimilé avec la transcription de Léopold Delisle* (Pretoria, S. Africa: Mousaion, 1965).The four citations appear on fol. 10v, 10v, 11r, and 13r. See also Léopold Delisle, *Le cabinet des manuscrits de la Bibliothèque National* (Paris, 1874), II, 526–28 (lines 43, 45, 47, and 59).

4. R. B. Thomson, "Jordanus de Nemore and the University of Toulouse," *The British Journal for the History of Science*, 1974, 7:163–65.

5. "qui cum parisiis in scientiis saecularibus et praecipue in Mathematicis manus habetur, libros duos admodum utiles unum de Ponderi et alium de Lineis datis dicitur edidisse," in F. Nicolae Triveti, *Annales sex regum Angliae* (edited by Thomas Hog, London, 1845), 211.

6. Peter Treutlein, "Der Traktat des Jordanus Nemorarius *de numeris datis*," *Abhandlungen zur Geschichte der Mathematik*, 1879, 2:129.

7. Maximilian Curtze, "Jordani Nemorarii Geometria vel De Triangulis Libri IV," *Mitteilungen des Coppernicus-Verein für Wissenschaft und Kunst zu Thorn*, 1887, 6:iii.

8. U. Chevalier, *Répertoire des sources historiques du moyen âge, bio-bibliographie* (Paris: Picard, 1905), II, 2647.

9. Moritz Cantor, *Vorlesungen über Geschichte der Mathematik* (New York: Johnson Reprint Corp., 1965 of the 1900 2d ed.), II, 57–60.

10. Marguerite Arons, *Saint Dominic's Successor* (trans. of *Un Animateur de la jeunesse au XIII siècle*; London: Blackfriars, 1955), passim but especially p. 26, where the author states that *De lineis* [sic] *datis* was printed at Nüremberg in 1537. Jordanus' work never had that title, apart from Trivet's remark, nor is there any record in such sources

as Maittaire's *Annales typographici* of its printing. The only extensive use of *De numeria datis* in the sixteenth century known to me is by Adam Riese in his *Coss*, which I discuss below.

11. N. Schreider, "The Beginnings of Algebra in Medieval Europe as Seen in the Treatise *de numeris datis* of Jordanus Nemorarius" (in Russian), *Ist. Math. Issled.*, 1959, *12*: 681.

12. Quoted by Curtze (n. 7 above), iii–vi, who did not accept Denifle's arguments.

13. E. A. Moody and M. Clagett, eds. *The Medieval Science of Weights* (Madison: Univ. of Wisconsin Press, 1952), 122–23.

14. For instance, Pierre Duhem, *Les Origines de la statique* (Paris, 1905), I, 389–93 et passim. See also P. L. Rose, *The Italian Renaissance of Mathematics* (Geneva: Librairie Droz, 1975), 81–82 et passim.

15. Marshall Clagett, *The Science of Mechanics in the Middle Ages* (Madison: Univ. of Wisconsin Press, 1961), 72.

16. Thomson (n. 1 above), passim, for manuscripts and printed editions. Only the *De elementis* is lacking a critical edition; editions of others, however, may well be improved.

17. Richard de Fournival mentions by name *Quemdam experimenta super algebra et abrakabala* (see Delisle, n. 3 above, 526–27). This title appears nowhere else in any literature that I have searched, including Thomson (n. 1 above). Since Richard de Fournival attributed it to Jordanus, I would like to think that it was a notebook in which lay the foundations for *De numeris datis*.

18. Michael S. Mahoney, "Another Look at Greek Geometrical Analysis," *Archive for History of Exact Sciences*, 1968, 5: 340. My thinking about Greek geometric analysis and algebra was influenced considerably by Mahoney's article, op. cit., 318–48, and by the work of Jacob Klein, *Greek Mathematical Thought and the Origin of Algebra* (Cambridge: M.I.T. Press, 1968). A more recent investigation deserves cautious reading: Jaakko Hintikka and Unto Remes, *The Method of Analysis: Its Geometrical Origin and Its General Significance* (Boston: D. Reidel Publ. Co., 1974).

19. *Pappi Alexandrini Collectionis quae supersunt*, ed. F. Hultsch (Berlin: Weidmann, 1877), II, 634–36; or Ivor Thomas, ed., *Selections Illustrating the History of Greek Mathematics* (Cambridge: Harvard Univ. Press, 1939), II, 596–98; or Sir Thomas L. Heath, *A History of Greek Mathematics* (Oxford: Clarendon Press, 1921), II, 400–01; but especially Mahoney (n. 18 above), 322–26, where one may find a closely reasoned argument for excising certain passages from the critical edition of Pappus (despite the remarks of Hintikka and Remes, n. 18 above, 48, n. 1). All quotations from Pappus' essay are from Mahoney's translation.

20. Francois Viète, *Introduction to the Analytical Art*, trans. J. W. Smith, in Klein (n. 18 above), 315–53.

21. Ibid., 320.   22. Ibid., 164.

23. Moody and Clagett (n. 13 above), 183.

24. Clagett (n. 15 above), 104–05, 107.

25. The text may be found in the Dresden Codex C 80m, quoted by Wolfgang Kaunzner, *Ueber Johannes Widmann von Eger, ein Beitrag zur Geschichte der Rechenkunst in ausgehenden Mittelalter* (München: Deutsches Museum, 1968), Reihe C, Quellentexte und Uebersetzungen, no. 7, pp. 139–42.

26. Two copies of the proposal are used as foresheets in Regiomontanus' copy of Ptolemy's *Geographia* (Basel, Oeffentl. Bibliotheca, Codex 0. IV. 32).

27. Moody and Clagett (n. 13 above), 151.

28. Thomas S. Kuhn, *The Structure of Scientific Revolutions*, 2d ed. (Chicago: Univ. of Chicago Press, 1970), 109.

29. Ibid., 11.

30. Three sources are quite useful here: A. C. Crombie, *Robert Grosseteste and the Origins of Experimental Science* 1100–1700 (Oxford: Clarendon Press, 1953), 57–60; Curtis Wilson, *William Heytesbury, Medieval Logic and the Rise of Mathematical Physics* (Madison: University of Wisconsin Press, 1960), 24–25; and especially, A. G. Molland, "The Geometrical Background to the 'Merton School', An Exploration into the Application of Mathematics to Natural Philosophy in the Fourteenth Centruy," *British Journal for History of Science*, 1968, *4*:109–11.

31. Moody and Clagett (n. 13 above), 159–61.

32. In his *Reason and Society in the Middle Ages* (Oxford: Clarendon Press, 1978), Alexander Murray does Leonardo an injustice. First, despite the good report he makes on *Liber abaci* in general, nowhere does he inform the reader that Fibonacci wrote the first comprehensive tract on elementary algebra, in chapter 13 of the book. Second, he implies that Fibonacci lacked a sense for algebra, where he describes how the Pisan solved a quadratic equation involving an irrational number (see p. 175). A further criticism is that Murray consistently exaggerates the role and function of computational arithmetic in late medieval society. Finally, he fails to distinguish between computation and theory of numbers, where he discusses the role of mathematics in the mastery of nature (see p. 206).

33. The entire tract was published by Boncompagni in *Trattati d'arithmetica*, II (Rome, 1857).

34. Ibid., 112.

35. L. C. Karpinski, *Robert of Chester's Latin Translation of the Algebra of al-Khowarizmi* (New York: Macmillan, 1915), passim. All references to al-Khwārizmī's *Liber algebre* are to this edition. The meaning of the word *algebra* has received close attention; see Solomon Gandz, "The Origin of the Term 'Algebra'," *American Mathematical*

Monthly, 1926, *33*:437–40, and G. Saliba, "The Meaning of al-jabr wa'l-muqābalah," *Centaurus*, 1972, *17*:189–204.

36. Frederic Rosen, *The Algebra of Mohammed ben Musa* (London, 1831), passim.

37. Karpinski (n. 35 above), 20–21.

38. Martin Levey, *The Algebra of Abū Kāmil, Kitab fi al-jabr wa'l-muqābala in a commentary by Mordecai Finzi* (Madison: Univ. of Wisconsin Press, 1966). See also L. C. Karpinski, "The Algebra of Abū Kāmil Shoja' ben Aslam," *Bibliotheca Mathematica*, 1912, ser. 3, *12*:40–55, and idem, "The Algebra of Abū Kāmil," *American Mathematical Monthly*, 1914, *21*:37–48.

39. George Sarton, *Introduction to the History of Science* (Washington, DC: Carnegie Institution, 1931), II, 341. The MS is in Paris, Bibliothèque Nationale, Latin 7377A, fol. 71v–93v.

40. Levey (n. 38 above), 14.

41. Levey remarked in his introduction, "Finzi, who quoted Campanus and Nemorarius . . ." (ibid., 13). While references to Euclid could be found (e.g. p. 42), no references to the other two were discovered in Finzi's translation of abū Kāmil, the subject of Levey's work.

42. Wertheim, "Ueber die Lösung einiger Aufgaben im *Tractatus de numeris datis* des Jordanus Nemorarius," *Bibliotheca mathematica*, 1900, ser. 3, *1*:417.

43. Levey (n. 38 above), 217 ff.

44. See, for example, Cambridge University Library MS Ii. 1. 13, fol. 17r.

45. Baldassarre Boncompagni, *Scritti di Leonardo Pisano* (Rome, 1857), I, 446. All references to the *Liber abaci* are to this edition.

46. Ibid., 395.

47. Ibid., 1. Similarly, in the prologue to the *Liber quadratorum*, where his number theory is truly creative, he wrote "the question put to me pertains as much to geometry as to number" (Ibid., II, 253).

48. See Diophantus, *Arithmetica*, IV, 39, in Ivor Thomas, *Selections Illustrating the History of Greek Mathematics* (London: William Heinemann, 1957), II, 530.

49. See al-Khwārizmī, *Liber algebre*, 23, where Karpinski discusses *radix*. But see also the text where both *radix* and *latus* are used: "Hoc autem quadratum quod loco substantiae ponimus, et eius radicem scire volumus atque designare" (p. 76, lines 24–25). Libri's transcription of BN Cod. lat. 7377, which contains a translation of al-Khwārizmī's text, renders the foregoing thus: "Fit ergo illi superficies quadrata ignotorum laterum que est census quem et cuius radices scire volumus" (G. Libri, *Histoire des sciences mathématiques en Italie* [Paris, 1838], I, 258). Gerard of Cremona's translation of the same varies considerably: "Ponam censum tetragonum *a b g d*, cuius radicem *a b* multiplicabo in decem

dragmas, que sunt latus *b e* eunde proveniat superficies *a e*" (Boncompagni, "Della vita et delle opere di Gherardo Cremonense," *Atti dell' Accademia de' nouvi Lincei*, 1851, 4:419); this is Vatican MS Latin 4606. The Bodleian MS Lyell 52 has exactly the same reading as the Vatican MS.

50. See n. 30 above. The definitive analysis of the transmission of Euclid's *Elements* has been done by John Murdoch, in *Dictionary of Scientific Biography*, IV, 437–59.

51. H. Menge, ed., *Euclidis Data* (Leipzig: Teubner, 1896), 2.

52. See Sir Thomas L. Heath, *History of Greek Mathematics* (Oxford: Clarendon Press, 1921), I, 370–71.

53. See Vera Sanford, *The History and Significance of Certain Standard Problems in Algebra* (New York: Columbia Univ. Press, 1927), passim.

54. Barnabas B. Hughes, "*De Regulis Generalibus*: A Thirteenth-Century English Mathematical Tract on Problem Solving," *Viator*, 1980, *11*:209–24.

55. Levey (n. 38 above), 32.

56. For a thorough discussion of the Babylonian origins of algebra and its influence upon successors, see Solomon Gandz, "The Origin and Development of the Quadratic Equations in Babylonian, Greek and Early Arabic Algebra," *Osiris*, 1938, *3*:405–557. To this may be added Evart M. Bruins and M. Rutten, *Textes mathématiques de Suse* (Paris: Paul Geuthner, 1961).

57. Florian Cajori, *A History of Mathematical Notations* (LaSalle, Ill.: Open Court Publ. Co., 1928), I, 336; and Thomas L. Heath, *Euclid's Elements*, 2d ed. (New York: Dover, 1956), II, 288. See Solomon Gandz, "On the Origin of the Word 'Root'," *American Mathematical Monthly*, 1926, *33*:261–65, and "On the Origin of the Word 'Root.' Second Article," ibid., 1928, *35*:67–75, for a much fuller development of "root" from historical and philological sources.

58. The text may be found in Libri (n. 49 above), I, 304–71. Libri attributed the book to Abraham ibn Ezra (1089?–1167), who traveled throughout England, France, and Provence to spread the Muslim scientific viewpoint among the Jews (p. 304). Neither Sarton (n. 39 above, II, 187–89 and 206–08) nor Levey (*Dictionary of Scientific Biography*, IV, 502–03) accepts Libri's argument.

59. See Levey (n. 38 above), 76 and 186 ff., and Boncompagni (n. 45 above), I, 374 ff.

60. Boncompagni (n. 45 above), 387.

61. Ettore Carruccio, *Mathematics and Logic in History and in Contemporary Thought* (London: Faber and Faber, 1964), 170–71.

62. There may have been more copies of English origin; but the destruction of libraries there in the mid-sixteenth century would have included these.

63. *Tabulae codicum . . . in Bibl. Palat. Vindobon. asservatorum* (Wien, 1870), III, n. 4770.

64. Treutlein (n. 6 above), 127–66; and Maximilian Curtze, "Die Ausgabe von Jordanus' *De numeris datis* durch Prof. P. Treutlein in Karlsruhe," *Amtl. Organ der Kaiser Leop.-Carol. deutsche Akademie der Naturforscher. Leopoldina*, Ser. 1, 1882, *18*:26–31.

65. Curtze, "Commentar zu dem 'Tractatus *De numeris datis*' des Jordanus Nemorarius," *Zeitschrift für Mathematik und Physik, Hist.-lit. Abtg.*, 1891, *36*:1–23, 41–63, 81–95, 121–38.

66. Schreider, n. 11 above.

67. R. D. von Sterneck, "Zur Vervollständigung der Ausgaben des Jordanus Nemorarius: 'Tractatus de numeris datis'," *Monatshefte für Mathematik und Physik*, 1896, 7:165–79; and Wertheim (n. 42 above), 417.

68. The incomplete list in the catalogue was fleshed out with the aid of a handwritten analysis of the codex by M. R. James, courteously supplied by A. E. B. Owen.

69. "P varies" means that P (or whichever MS is said to vary) has the minority reading, which was rejected as incorrect.

70. D, fol. 230r. For the entire theorem, sixty percent of which is illegible in the MS, see Curtze (n. 65 above), 14n*.

71. C, fol. 381v.   72. See n. 25 above.

73. See Bruno Berlet, *Adam Riese, sein Leben, seine Rechenbücher und sein Art zu Rechen* (Leipzig, Frankfurt: Kesselringsche Hofbuchhandlung Berla, 1892), 27–62, esp. 32 n. 4.

The Critical Edition

These sigla identify the texts used
to prepare the critical edition:

A Milano, Bibl. Ambrosiana, *Cod. Lat. D 186 Inf.*

B Basel, Oeffent. Bibliothek, *Cod. Lat. F. II. 33.*

D Dresden, Sächs. Landesbibliothek, *Cod. Lat. Db 86.*

F Firenze, Bibl. Nazionale, *Conv. Soppr. J. V. 18.*

J Krakow, Bibl. Jagiellonska, *MS 1924.*

M Paris, Bibl. Mazarine, *Cod. Lat. Nr. 3642.*

O Oxford, Bodleian Library, *MS Auct. F. 5. 28.*

P Paris, B. N., *Cod. Lat. 8680 A.*

R Paris, B. N., *Cod. Lat. 11863.*

S Paris, B. N., *Cod. Lat. 11885.*

V Vatican, Bibl. Apostolica, *Cod. Ottobon. Lat. 2120.*

W Wien, Nationalbibliothek, *Cod. Lat. Nr. 4770.*

X Wien, Nationalbibliothek, *Cod. Lat. Nr. 5303.*

De datis numeris

Numerus datus est, cuius quantitas nota est.
Numerus ad alium datus est cum ipsius
ad illum est proportio data.
Data est autem proportio cum ipsius
denominatio est cognita.

I-1.

SI NUMERUS DATUS IN DUO DIVIDATUR QUORUM DIFFERENTIA DATA, ERIT
UTRUMQUE EORUM DATUM. Etenim minor portio et differentia faciunt
maiorem. Tunc minor portio cum sibi equali et cum differentia facit
totum. Sublata ergo differentia de toto, remanebit duplum minoris
5 datum. Quo diviso, erit minor portio data. Sicut et maior. Verbi
gratia: x dividatur in duo. Quorum differentia duo. Qui si auffera-
tur de x, reliquitur viii cuius medietas est iiii et ipse est minor
portio. Altera, vi.

I-2.

SI NUMERUS DATUS DIVIDATUR PER QUOTLIBET, QUORUM CONTINUAE
DIFFERENTIAE DATA FUERINT, QUODLIBET EORUM DATUM ERIT. Datus
numerus sit a qui dividatur in b, c, d, e sitque e minimus. Et
quia eorum sunt continuae differentiae data, singulorum ad e da-
5 tae erunt differentiae. Sit igitur f differentia b ad e, et g, h
differentiae c ad e et d ad e. Et quia e cum singulis illorum facit singula istorum, manifestum est quod triplum e cum f g h
facit illos tres. Quadruplum ergo e cum f g h facit a. Hiis ergo
demptis de a remanebit quadruplum e datum. Quare e datum erit, et

10 per additionem differentiarum, erunt reliqua data. Hoc opus est.
Verbi gratia: xl dividatur per iiii quorum per ordinem differentiae sint iiii, iii, ii. Differentia ergo primi ad ultimum ix, et
secundi ad illum v, et tertii ad eundem duo. Quae simul faciunt xvi.
Quibus demptis de xl remanebunt xxiiii, quorum quarta est vi et hoc
15 erit minimus quatuor. Additis autem ix, v et duobus proveniunt
ceteri tres, vii, xi, xv.

    I-3.

DATO NUMERO PER DUO DIVISO SI, QUOD EX DUCTU UNIUS IN AL-
TERUM PRODUCITUR, DATUM FUERIT, ET UTRUMQUE EORUM DATUM ESSE
NECESSE EST. Sit numerus datus abc divisus in ab et c, atque ex
ab in c fiat d datus, itemque ex abc in se fiat e. Sumatur ita-
5 que quadruplum d, qui sit f, quo dempto de e remaneat g, et ipse
erit quadratum differentiae ab ad c. Extrahatur ergo radix
g et sit b, eritque b differentia ab ad c. Cumque sit b datum,
erit et c et ab datum. Huius operatio facile constabit hoc modo.
Verbi gratia: Sit x divisus in numeros duos, atque ex ductu
10 unius eorum in alium fiat xxi, cuius quadruplum et ipsum est
lxxxiiii, tollatur de quadrato x hoc est c, et remanent xvi cuius
radix extrahatur, quae erit iiii et ipse est differentia, ipsa
quae tollatur de x et reliquum, hoc est vi, dimidietur. Eritque medietas iii, et ipse est minor portio et maior vii.

    I-4.

SI NUMERUS DATUS FUERIT IN DUO DIVISUS, QUORUM QUADRATA
PARITER ACCEPTA SINT DATA, ERIT UTRUMQUE DATUM. Modo praemisso si enim g fuerit notus, erit et e notus, qui est duplum unius in alterum. Subtractoque e de g remanebit h, quad-
5 ratum differentiae, cuius radix extracta cum sit nota
erunt omnia data. Opus idem. Divisus quippe sit x in duo,
quorum quadrata sint lviii, quo sublato de c remanebunt
xlii, et ipse auferatur de lviii remanebunt xvi, radix cuius
est quatuor, et ipsa est differentia portionum, quae fient
10 vii et tres, ut prius.

    I-5.

SI NUMERUS IN DUO DIVIDATUR, QUORUM DIFFERENTIA DATA,
ATQUE EX DUCTU UNIUS IN RELIQUUM PROVENERIT NUMERUS DATUS,

NUMERUM QUOQUE DIVISUM DATUM ESSE CONVENIET. Maneat superior
dispositio, et *b*, differentia portionum, sit datus, et simil-
5 iter *d*, qui est productus ex eis, cuius duplum *e* est. Et *e*
duplicato addatur *h*, qui est quadratum differentiae, et compositus sit *f*, qui erit quadratus *abc* et datus, quare et *abc*
datus est. Verbi gratia: Differentia portionum sit vi, et ex ipsis
proveniant xvi, cuius duplum xxxii, illius quoque duplum lxiiii.
10 Huic addatur xxxvi, scilicet quadratum vi, et fient *c*, cuius
radix extracta erit x, numerus divisus in viii et duo.

I-6.

SI VERO DIFFERENTIA DATA FUERIT ET QUADRATA EORUM CONIUNCTIM
DATA, NUMERUS ETIAM TOTUS DATUS ERIT. Quadrata eorum coniuncta er-
ant *g*, qui sit datus, de quo tollatur *h* quadratus differentiae,
similiter datus, et remanebit *e* datus qui est duplus unius in al-
5 terum, additoque *e* ad *g* fiet *f*, quadratus totius. Extracta ergo
radice *f* erit totus *abc* datus. Verbi gratia: lxviii sint duo
quadrata a quibus tollantur xxxvi qui est quadratus differentiae,
et remanebunt xxxiii qui est duplum unius in alterum. Coniuctis
itaque lxviii et xxxii provenient *c*. Huius radix est x, et ipse
10 divisus in viii et ii.

I-7.

SI DIVIDATUR NUMERUS IN DUO, QUORUM ALTERUM TANTUM DATUM,
EX NON DATO AUTEM IN SE ET IN DATUM PROVENERIT NUMERUS DATUS,
ERIT ET NUMERUS QUI DIVISUS FUERAT DATUS. Sit numerus divisus
in *a* et in *b* sitque *b* datus atque ex *a* in se et in *b*, hoc est
5 in totum *ab*, proveniat *d* qui sit datus. Addatur autem *c* ad
ab et ipse sit equalis *a*, ut sit totus *abc* divisus in *ab* et *c*.
Quia igitur ex *ab* in *c* fit *d* datus, atque differentia *ab* ad *c*,
scilicet *b*, est datus, erit *abc* et *c* datus, similiter et *a* et
ab. Huius operatio est verbi gratia. Sit vi unum dividentium,
10 et ex reliquo in se et in vi fiant xl quorum duplum id est
lxxx duplicentur, et erunt clx, quibus addatur quadratum vi hoc
est xxxvi, et fient cxcvi, cuius radix est xiiii, de quo sublatis vi et reliquo mediato fient iiii, qui est reliquum. Eritque totus divisus x, coniunctis iiii et vi.

I-8.

SI NUMERUS DATUS IN DUO DIVIDATUR, ET EX DUCTU TOTIUS IN
DIFFERENTIAM ET MINORIS DIVIDENTIUM IN SE PROVENERIT NUMERUS DA-
TUS, ERIT ET UTRUMQUE ILLORUM DATUM. Illa enim coniuncta sunt
tamquam quadratum maioris numeri. Extracta igitur radice illius
5 habebis maius dividentium, et ita reliquum. Verbi gratia: Dividatur x in duo, et ex ductu ipsius in differentiam et minoris
portionum in se fiant lxiiii. Radix cuius est viii, qui erit
maior portio, et ii minor.

    I-9.

SI VERO EX DUCTU TOTIUS IN DIFFERENTIAM ET MAIORIS DIVI-
DENTIUM IN SE FIAT NUMERUS DATUS, UTRUMQUE ETIAM DATUM ERIT.
Esto *ab* divisus in *a* et *b*, quorum differentia *c*, atque ex *ab*
in *c* fiat *d*, et ex *a*, qui est maior, in se fiat *e*, eritque to-
5 tus *de* datus. Sed et *ab* in se faciat *f*. Quare totus *def* datus est. Sed quia *abc* duplus est *a*, erit *df*, quod fit ex *ab* in
duplum *a*. Erit ergo *df*, quod fit ex duplo *ab* in *a*. Sic igitur *def* erit, quod provenit ex *a* in se et in duplum *ab*. Cumque *def* sit datum sed et duplum *ab*, erit et *a* datus et ideo *b*.
10 Verbi gratia: x in differentiam portionum et maior portio in se
faciant lvi, quibus iungantur c, et erunt clvi quorum duplum
hoc est cccxii duplicetur et fient dcxxiiii quibus addatur
quadratum xx, qui est duplum x, et fient mxxiiii, huius radix
xxxii de quo tollatur xx et remanebunt xii cuius dimidium vi,
15 et ipse est maior portionum x, et reliqua iiii.

    I-10.

QUOD SI QUADRATA DIVIDENTIUM AMBO CUM EO, QUOD EX TOTO IN
DIFFERENTIAM, FECERINT NUMERUM DATUM, QUODLIBET EORUM DATUM
ESSE NECESSE EST. Omnia enim haec sunt tamquam duplum quadrati
maioris dividentis. Dimidientur itaque et dimidii extrahatur ra-
5 dix, et habebitur maior portio. Verbi gratia: Diviso x quadrata portionum et quod fit ex x in eorum differentiam, omnia sint
xcviii cuius medietas est xlix cuius radix est vii et ipse est
maius dividens, minor vero est iii.

    I-11.

SI ITEM QUOD FIT EX TOTO IN DIFFERENTIAM CUM EO, QUOD EX
UNO DIVIDENTIUM IN RELIQUUM PRODUCITUR, FUERIT DATUM, ERUNT
SINGULA EORUM DATA. Cum sit autem totum ex differentia et duplo

minoris dividentium compositum, tantum erit totum in se,
5 quantum semel in differentiam et minor portio bis in ipsum.
Sed minor in totum, tantum est quantum in maiorem et in se. Si
ergo quod fit ex toto in differentiam cum eo, quod ex minore
dividentium in reliquum tollantur de quadrato totius, remanebit
quod fit ex minore in se et in totum datum. Sic ergo ex prae-
10 missis et ipsum datum erit, et reliquum. Operis executio verbi
gratia. Quod fit ex x in differentiam, cum eo quod ex uno dividentium in alterum, faciat lxxxix; quo sublato de c remanent xi
cuius duplum dupletur, et fient xliiii quae cum c sunt c et xliiii
quorum radix est xii. Huius ad x differentia est duo quorum medi-
15 etas est unum, et ipse minus dividens et maius ix.

    I-12.

SI NUMERO DATO PER DUO DIVISO QUADRATA IPSORUM CUM QUAD-
RATO DIFFERENTIAE FUERINT DATAE, UTRUMQUE EORUM DATUM ERIT.
Detractis siquidem omnibus hiis de quadrato totius remanebit minus duplo unius in alterum quantum est quadratum differentiae.
5 Quare minus duobus quadratis dividentium duplo eiusdem quadrati;
minus ergo toto detracto, eius triplo. Cum ergo ipsum residuum
de detracto sublatum fuerit, reliqui sumatur tertia, cuius radix
differentia erit et data. Omnia ergo data. Verbi gratia: Diviso
x in duo sunt quadrata eorum cum quadrato differentiae lvi qui
10 tollatur de c, et remanebunt xliiii et hic auferatur de lvi et
reliquuntur xii quorum tertia est iiii. Huius radix est duo
et ipse est differentia portionum. Maior itaque erit vi et minor
iiii.

    I-13.

SI VERO QUOD FIT EX DUCTU ALTERIUS IN ALTERUM CUM QUADRATO
DIFFERENTIAE FUERIT DATUM, DATUM ERIT ET UTRUMQUE IPSORUM.
Totum duplicetur, et fient tamquam duo quadrata et quadratum
differentiae, quae quoniam sunt data, sunt etiam et quae proponimus.
5 Verbi gratia: Ductum unius in alterum cum quadrato differentiae
sint xxviii quae duplata faciunt lvi quae sunt tria quadrata ut supra,
et cetera eodem modo.

    I-14.

SI NUMERUS DATUS IN DUO DIVIDATUR, ET QUADRATO MINORIS
ET QUADRATO MAIORIS DETRACTO RELIQUUM DATUM FUERIT, ERUNT ET

IPSA DATA. Illo enim detracto de quadrato totius relinquitur
quadratum minoris bis et quod fit ex ipso in reliquum bis. Si
5 ipsum igitur dimidietur, proveniet medietas. Quadratum minoris
semel et quod fit ex ipso in maius, et hoc tantum est quantum si
ducatur totum in minorem portionum. Dividatur ergo per totum,
et exhibit minus dividentium. Modus operationis. Verbi gratia:
Divisus sit item x in duo, et quadrato minoris detracto de quadrato
10 maioris relinquatur lxxx, quod minuit xx de c, cuius medietas
est x quo diviso per x exit unum, et ipsum est minus dividentium
et ix maius.

I-15.

NUMERO DATO PER DUO DIVISO QUADRATIS EORUMDEM DIFFERENTIA
ADDITA SI NUMERUM DATUM FECERINT, SINGULA EORUM DATA ERUNT.
Hoc de quadrata totius si detractum fuerit, manifestum est relinqui
minus detracto, quantum est differentia bis cum quadrato ipsius,
5 hoc est, quod fit ex ipso in se et in binarium, qui est datus.
Quare et differentia data erit. Verbi gratia: Divisus sit item
x per duo, quorum quadrata addita differentia fiant xii. Ista
tollantur de c, remanebunt xxxviii. Haec si auferantur de lxii,
relinquuntur xxiiii, qui fit ex ductu differentiae in se et in
10 binarium. Ergo dupletur duplum et fient xcvi quibus addantur
iiii quod est quadratum binarii, et fiunt c. Huius radix est x,
de quo subtractis duobus reliqui dimidium, hoc est iiii, erit
differentia. Sunt ergo dividentia vii et iii.

I-16.

QUOD SI ADDITA EADEM DIFFERENTIA EI, QUOD FIT EX UNO IN
RELIQUUM, FUERIT TOTUM DATUM, DATUM ERIT SINGULUM EORUM. Sit
ab numerus divisus et quod fit ex *a* in *b* addita differentia
sit *c*, et ipsum duplicatum sit *d*. Quadratum autem totius sit
5 *e*, de quo detracto *d* remaneat *f*. Qui si fuerit minor *d*, videatur quanto. Quia si iiii, differentia erit duo. Si tribus,
differentia erit tres vel unum. Sed hoc determinari non potest. Si equales fuerint *d* et *f*, differentia erit iiii. Si vero
f excedit *d*, videatur quanto sitque *g*. Eritque *g*, quod fit ex
10 ductu illius, quo differentia excedit duplum binarii in se et
in illud duplum. Quare et ipsum datum erit, et tota differentia *a* ad *b* data. Huius opus est huiusmodi verbi gratia. Divi-

datur ix in duo, et ex ductu unius in alterum addita differentia fiant xxi, cuius duplum quod est xlii, tollatur de lxxxi, et
15 remanebunt xxxix, quae minuunt iii de xlii. Potest ergo esse
differentia unum et iii et utrumque contingit. Unum erit. Si
divisus fuerit ix in v et iiii, et v in iiii addito uno, faciunt xxi. Tria erunt divisio ix in vi et iii, et si hoc iii in vi
additis iii faciunt xxi. In hoc ergo error incidit. Item diviso
20 ix proveniant xix cuius duplum xxxviii. Hoc si auferatur de
lxxxi, relinquentur xliii, qui illum excedit v. Huius duplum
dupletur et fiunt xx. Huic quadratum additur iiii qui est duplum
duorum, et erunt xxxvi, cuius radix vi, de quo detracto iiii reliqui dimidium erit unum, et hoc cum iiii facit v, et ipse est
25 differentia portionum, quae sunt vii et duo.

I-17.

DATO NUMERO IN DUO DIVISO SI, QUOD FIT EX UNO IN RELI-
QUUM, PER DIFFERENTIAM DIVIDATUR, ET QUOD EXIERIT, FUERIT DA-
TUM, ERIT ET UTRUMQUE DIVIDENTIUM DATUM. Quia enim, quod fit
ex uno in reliquum quantum continetur in quadrato totius cum
5 quadrato differentiae, erit ut differentia ducta in se et in
quadruplum dati numeri qui exierat, faciat quadratum numeri
divisi. Data ergo erit et differentia. Verbi gratia: Dividatur x in duo, et quod fit ex uno in reliquum, diviso per differentiam exeat xii. Huius quadruplum est xlviii. Dupli igi-
10 tur c sumatur duplum, huic addatur quadratum xlviii, quod est
iī.ccciiii, et fiant iī.dcciiii cuius radix est lii de quo subtracto xlviii reliqui medietas est duo, et ipse est differentia portionum.

I-18.

SI VERO QUADRATA EORUMDEM CONIUNCTA PER DIFFERENTIAM
DIVIDANTUR, ET QUOD EXIERIT FUERIT DATUM, ET EORUM QUODLIBET
DATUM ERIT. Sit datus numerus *ab* divisus in *a* et in *b*, quorum quadrata sint *c*, et differentia eorum *d*, cuius quadratum
5 *e*, et quadratum totius *f*. Diviso ergo *c* per *d* exeat *g*, cuius
duplum sit *hl*, qui erit datus. Et quia quadrata *e* et *f* sunt
duplum *c*, erit, ut *d* in *hl* faciat *ef*. Sit autem *l* equale *d*.
Et quia *l* in se faciat *e*, tunc *l* in *h* faciat *f* qui est notus.
Et quia *hl* est notus, erit et *l* et *h* datus, sicque *d* et omnia.

10 Verbi gratia: Divisus sit x in duo, quorum quadrata divisa per
differentiam reddant xxvi cuius duplum est lii huius quadratum īīdcc et iiii. Ab hoc tollatur c quater, et remanebunt
īīcccii, cuius radix est xlviii, hic detrahatur a lii et
reliqui medietas, quae est duo, est differentia portionum.

> I-19.

SI NUMERUS DATUS IN DUO DIVIDATUR, UNOQUE EORUM PER
RELIQUUM DIVISO EXIERIT NUMERUS DATUS, ET IPSA DATA ESSE
OSTENDETUR. Dividatur *a* per *b*, et exeat *c* datum, cui addito
uno fiat *d*. Et quia *b* in *c* facit *a*, tunc in *d* faciet *ab*.
5 Dividatur ergo *ab* per *d*, et exibit *b*. Verbi gratia: Dividatur x in duo, et uno diviso per reliquum fiat iiii, cui
addito uno fient v, per quem divisus x facit duo, qui est
una portio.

> I-20.

QUOD SI UTRUMQUE PER RELIQUUM DIVIDATUR, ET QUAE EXIERINT
CONIUNCTA DATUM QUID FECERINT, ERUNT SIMILITER ET IPSA DATA.
Dividatur *a* per *b* et exeat *c*, et *b* per *a* et prodeat *d*. Singu-
lae etiam unitates addantur *c* et *d* et fiant *e* et *f*, atque ex
5 *a* in *b* fiat *g*. Quia igitur ex *a* in *f* facit *ab*, atque *b* in *e*
facit *ab*, erit *e* ad *f* sicut *a* ad *b*. Quare *e* ad *f* sicut *ab* ad
b, et permutatim *ab* ad *ef* sicut *b* ad *f*. Et quia *a* in *b* et in
f facit *g* et *ab*, erit *g* ad *ab* sicut *b* ad *f*. Quare *g* ad *ab* sicut *ab* ad *ef*. Quadratum igitur *ab* dividatur per *ef* quod est
10 datum, et exibit *g* datum. Erit ergo et *a* et *b* datum.
Opus ergo breve. Verbi gratia: Dividatur x in duo, quorum
utrumque per reliquum dividatur, et quod exit totum sit duo
et sexta. Quibus addatur duo et fient iiii et sexta, per
quod dividatur c, et exibunt xxiiii, et ipsum fit ex una
15 portione in reliquam. Quater ergo, ut solet, detrahatur
de c, et remanebunt iiii cuius radix est duo; et ipse est
differentia dividentium quae sunt vi et iiii.

> I-21α.

DATO NUMERO IN DUO DIVISO SI SECUNDUM UTRUMQUE EORUM
QUILIBET NUMERUS DATUS DIVIDATUR, ET QUAE EXIERINT FECERINT
NUMERUM DATUM, EORUM QUODLIBET DATUM ERIT. Cum *c* numerus da-
tus per *a* et *b* dividatur, et exeat coniunctim *de* datum, ita-

5 que *c* per *ab* divisus reddat *f*. Et quia, quod fit ex *f* in
quadratum *ab*, quod sit *g*, est quantum quod fit ex *de* in productum ex *a* in *b*, quod sit *h*. Itemque quod fit ex *f* in *g* est
quantum, quod est ex *ab* in *c*. Ideo ducatur *ab* in *c*, et productum dividatur per *de* et exhibit *h* datum, quare et *a* et *b*
10 datum erit. Verbi gratia: Diviso x in duo per utrumque dividatur xl, et exeat xxv. Ducatur autem x in xl, et productum
dividatur per xxv, et exibit xvi, et ipse fiet ex uno dividentium
in reliquum.

I-21β.

DATO NUMERO IN DUO DIVISO, ET ALTERO DIVIDENTIUM PER
DATUM NUMERUM MULTIPLICATO ET RELIQUO PER PRODUCTUM, SI QUOD
PERVENIT, DATUM FUERIT, UTRUMQUE DIVIDENTIUM DATUM ERIT. Sint
dividentia *a*, *b* atque *a* multiplicetur per *c* datum numerum et
5 proveniat *d*, atque *d* in *b* faciat *e* datum. Quia ergo multiplicato sic secundum terminum facit quantum si *a* in *b* et *c*, et
productum ducatur. Tunc *e* dividatur per *c* et exibit datum
quod facit ex *a* in *b*. Quadrata data etiam erit *a* et *b*, quia
ab datus est. Verbi gratia: Diviso x, ducatur v in alterum
10 dividentium et reliquum in productum et faciant cv. Hic totus
dividatur per v et exibit xxi qui fiet ex una portionum x
ductum in reliquum, ut ergo consuetudo est, xxi quater tollatur de c et remanebunt xvi. Huius radix extracta est iiii
qui est differentia dividentium et ipsae sunt vii et iii.

I-22α.

SI VERO EX DUCTU UNIUS IN RELIQUUM PROVENERIT ALIQUOD
DATUM, UTRUMQUE EORUM DATUM ESSE CONVENIET. Fiat enim *f* ex *d*
in *e*, atque ex *a* in *b* fiat *h*. Quia igitur *a* in *b* et *d* fiunt
h et *c*, erit *c* ad *h* sicut *d* ad *b*. Itemque ex *e* in *b* et *d* fiunt
5 *c* et *f*, erit *f* ad *c* sicut *d* ad *b* et sicut *c* ad *h*. Si ergo quod
fit ex *c* in se dividatur per *f*, exibit *h*. Verbi gratia: Diviso
xl per portiones x, et uno in aliud ducto fiant c per quod
si dividatur quod fit ex xl in se, exibunt xvi ut prius.

I-22β.

SI VERO PRODUCTUM PER RELIQUUM DIVIDENTIUM DIVIDATUR ET
QUOD EXIERIT DATUM FUERIT, ERUNT SINGULA DATA. Itaque *d* dividatur per *b* et exeat *e* datum. Quia igitur *a* ad *e* sicut *b* ad

c erunt ut *ab* ad *ec* sicut *a* ad *e*. Itaque *ab* ducatur in *e* et
5 dividatur per *ec*, et exibit *a* datum, et sic *b* erit datum.
Verbi gratia: x per duo diviso et altero per v multiplicato,
et producto per alterum diviso exeat vii et dimidium. Itaque
x ducatur in v et fiet l, qui dividatur per xii et dimidium
et exibunt iiii qui erunt una portio.

    I-23.

QUOD SI UNUM EORUM PER RELIQUUM DIVIDATUR, ET QUOD
PROVENERIT DATUM FUERIT, SINGULUM EORUM DATUM ERIT. Esto
ut prius quod *c* dividatur per *a* et *b* et proveniant *d* et *e*,
atque *d* dividatur per *e* et exeat *f* datum. Et quodlibet quod
5 fit ex *a* in *d*, est quantum quod ex *b* in *e*, scilicet *c*, erit
a ad *b* sicut *e* ad *d*. Diviso ergo *d* per *e* tantum exit quantum si *b* dividatur per *a*, quod cum datum sit, palam quod omnia data esse constat. Verbi gratia: Diviso x in duo per
utrumque dividatur xl, et eorum quae exeunt uno diviso per al-
10 terum exeat quarta. Erunt ergo portiones x duo et viii.

    I-23*.

QUOD SI ALTERO DIVIDENTIUM PER DATUM NUMERUS DIVISO,
QUOD EXIERIT PER RELIQUUM DIVIDATUR, QUODQUE TANDEM PROVENERIT
DATUM FUERIT, ERUNT ET IPSA DATA. Per *c* datum numerum dividatur *a* et exeat *d*, qui dividatur per *b* et proveniat *e*. Quia
5 igitur *b* in *e* facit *d* et *d* in *c* facit *a*. Si *c* ducitur in *e* et
fiat *f*, erit *f* datum atque *b* in *f* facit *a* atque *b* et *a* datum.
Verbi gratia: Diviso x in duo alterum per v et quod provenerit diviso per reliquum et proveniat quinta. Quae multipli-
catur per v et fiet unum. Quod erit pars unius portionum denomi-
10 nata. Et reliquam autem addatur unum et erunt duo per quae
dividantur x et exibunt v, quod erit unum dividentium.

    I-24.

NUMERO DATO PER DUO DIVISO SI ALTERUM PER ALTERUM
DIVIDATUR ET ILLIUS, QUOD EXIERIT, QUOTALIBET PARS DIVISO
ADDATUR UT TOTUM DATUM SIT, UTRUMQUE EORUM DATUM ERIT.
Dividatur *a* per *b*, et quod exierit sit *c*, cuius medietas
5 addatur *a*, ut fiat *ad* datum. Perpende igitur, utrum sit
maius *ab* an *ad*. Sitque ut *ab*. Et maiori semper addatur to-

ta pars unius quota pars *c* additur *a*, ut *abe*, sitque *e* dimidium unius. Quia igitur *d* in *b* bis facit *a*, et in *e* bis facit se ipsum, erit ut in *eb* bis ductum faciat totum *ad*. Posito
10 ergo quod *g* sit differentia *abe* super *ad*. Itaque *d* bis in
se et in *g* facit *ad*. Semel ergo ductum in se et in *g* faciet
dimidium *ad*, quod cum sit datum, etiam *g* datum erit et *d* et
a datum. Quod si *ad* maius fuerit, et tunc *b* in se et in *g*
faciet dimidium *ad*, et ita similiter omnia erunt nota. Item
15 si *ad* et *ab* sunt equalia, erit ut *b* in se et in dimidium unius
quod est *e*, faciat dimidium *ad*, et sic eadem ratio erit. Scien-
dum etiam hoc opus triplex contingit. Verbi gratia: Diviso x
in duo ponatur alterum per alterum dividi, et medietas eius quod
prodierit addatur diviso, ut sit totum iiii et tertia, cuius
20 ad x et dimidium differentia est vi et sexta. Itaque iiii et
tertia dimidietur, et dimidium ut solet quadruplicetur, cui addatur quod fit ex vi et sexta in se, et erunt xlvi et ii
tertiae et trigesima sexta, cuius radix est vi et ii tertiae
et sexta. Ab hoc tollatur vi et sexta, et relinquuntur ii
25 tertiae, cuius dimidium tertia est, qua sublata de iiii et
tertia remanebunt iiii, et ipse est altera portionum.

I-25.

DATO NUMERO IN DUO DIVISO ET ALTERO DIVIDENTIUM PER DATUM
NUMERUM MULTIPLICATO, PRODUCTO QUOQUE PER ALTERUM DIVISO,
SI EIUS QUOD EXIERIT QUOTACUMQUE PARS PRODUCTO ADDITA TOTUM
FECERIT DATUM, SINGULA DATA ESSE NECESSE EST. Ut si *a* per *c* datum
5 numerum multiplicetur, et proveniat *d* qui dividatur per *b*, et exeat *e* cuius pars quotalibet sit *f* quae addatur *d*, ut fiat *df* numerus datus. Qui totus dividatur per *c* et prodeat *gh*. Sitque *g*
equalis *a*. Eritque *h* qui multiplicatus per *c* faciat *f*, et quia
c ad *b* sicut *e* ad *a*, quia *c* in *h* facit *f*, erit ut *b* in *h* faciat
10 totam partem *a*, et quia totum *gh* datum erit et *ah*, et ob hoc *a*
et *b* datum. Verbi gratia: Dividatur x in duo quorum alterum per
v multiplicetur et producto per reliquum diviso medietas eius,
quod exierit, eodem producto addatur, ut sit totum l quod dividatur per v, et exibunt x. Restatque nunc opus praemissae ubi
15 incidit equalitas. Medietas itaque x quadruplicetur, et fient

xx cui addatur quadratum dimidii, hoc est quarta, et erunt xx et
quarta cuius radix est iiii et medietas unius, de quo sublato dimidio et reliquo dimidiato exibunt duo, et ipse est unum dividentium.

I-26.

SI NUMERUS DATUS IN DUO DIVIDATUR, QUAE PER SINGULOS DATOS
NUMEROS DIVIDANTUR, ET QUAE PROVENERINT CONIUNCTA DATUM
NUMERUM CONSTITUANT, QUEMLIBET EORUM DATUM ESSE CONVENIET.

A per c et b per d datos numeros dividantur, et exeant e et f, sitque ef
5 datum, maior autem numerorum c et d sit c, cuius ad d differentia
sit g. Ducatur itaque d in ef et fiet mn, et n sit equalis b,
sed quo m minus est a, sit l. Dividaturque l per g et exibit e
datum. Quare et a et b data. Verbi gratia: Ut solet x in duo
secatur quorum alterum dividatur per iii et alterum per duo, et
10 exeant quatuor in quae ducantur duo et fient octo, et reliqua
duo de x dividatur per unum quod est differentia trium ad duo
et exeant duo in quae ducantur tria, et fient vi quae est una
portio.

I-27.

SI VERO ALTERUM IN ALTERUM DUCATUR, FUERITQUE PRODUCTUM
DATUM, OMNIA DATA ESSE DEMONSTRABITUR. Ducatur e in f, et fiat
g datum. Ducaturque c in g et fiat h quod tantum erit quantum
si f ducatur in productum ex c in e, hoc est in a. Ducatur item
5 d in h et producatur l quod etiam tantum erit quantum si a ducatur in productum ex d in f, hoc est in b quod cum datum, erit et
a et b datum. Verbi gratia: Diviso ergo x in duo, unumque per
iiii, alterum per duo partiatur, et quae exierint unum ductum in
alterum faciat duo. Quae duo multiplicentur per iiii et productum
10 per duo, et exibunt xvi, et ipse erit qui fit ex ductu unius dividentium in reliquum, quae ex hoc constabit esse viii et duo.

I-28α.

DIVIDATUR ALTERUM PER ALTERUM, TUNC SI EXIERIT QUOD
DATUM, OMNIA DATA ESSE CONSEQUETUR. Dividatur e per f et
exeat h datum. Dividatur item h per d et prodeat k, et k
multiplicetur per c, et fiet l. Quia igitur f in h facit e,
5 tunc b in k faciet e, et sic b in l producat a. Si ergo a
dividatur per b, exibit l, quod cum sit datum, erit a et b

datum. Verbi gratia: x dividatur in duo, et quarta unius dividatur per dimidium alterius, et exeat tertia, cuius dimidium quadruplicetur, exibuntque duae tertiae. Dividatur ergo
10 ut solet x per unum et duas tertias, et provenient ⟨vi⟩, et ipse
est una portio x.

I-28β.

DATO NUMERO PER DUO DIVISO, SI SECUNDUM UTRUMQUE EO-
RUM QUILIBET IDEM NUMERUS DATUS DIVIDATUR ET QUAE EXIERUNT
FECERINT UNUM DATUM IPSORUM QUODLIBET DATUM ESSE CONVENIET.
c numerus datus per a et b dividatur et exeat coniunctim *de* da-
5 tum. Itaque c per ab divisus reddat f. Et quia quod fit ex f in
quadratum ab quod sit g est ⟨quantum⟩ quod sit ex *de* in id quod
fit ex a in b quod sit h. Itaque quod fit ex f in g est quantum ex ab in c. Ideoque ducatur ab in c et productum dividatur
per *de*, et exibit h datum. Quare et a et b datum erit. Verbi
10 gratia: Diviso x in duo per uterque dividatur xl et exeat xxv.
Ducatur item x in xl et productum dividatur per xxv, et exibit xvi.
Et ipse est qui fit ex uno dividentium in reliquum.

I-29.

SI NUMERUS DATUS IN DUO DIVIDATUR, ATQUE QUOD FIT EX
TOTO IN ALTERUM EQUALE SIT QUADRATO ALTERIUS, ERIT UTRUMQUE
DATUM AD PROXIMUM. Sit ut ex ab in b sit quantum ex a in se.
Et quia quod ex ab in se est quantum quod ex ab in a et in b,
5 erit etiam quantum quod ex a in se et in ab. Cumque sit ab
datum, et a et b datum. Verbi gratia: x dividatur in duo ita
quod ex uno in alterum fit, quantum reliquum in se. Itaque x in
se facit c, cuius dupli duplum sumatur, et erunt cccc; huic
addatur, ut solet, quadratum x et erunt d, cuius radix extra-
10 hatur ad proximum, et erit xxii et tertia, de quo tollatur x,
et reliqui medietas erit vi et sexta, et ipsum erit maior por-
cionum quae ducenda est in se.

EXPLICIT LIBER PRIMUS XXIX CONTINENS PROPORTIONES.

Incipit liber secundus

II-1.

SI FUERINT QUATUOR NUMERI PROPORTIONALES, ET TRES EORUM
DATI FUERINT, ET QUARTUS DATUS ERIT. Facta enim altera multiplicatione idem numerus producitur. Sumptis ergo alternatim,
quoniam duo sunt dati, alter in alterum ducatur, et productus
5 per unum reliquorum, qui datus est, dividatur, et exibit reliquus datus qui prius fuerat non datus. Verbi gratia: Sint
xx ad aliquem sicut v ad iiii. Quia igitur ducendus est antecedens datus in consequentem alterius datum, ducatur xx in iiii
et fient lxxx qui dividatur per v et exibunt xvi qui erit con-
10 sequens xx prius non datus.

II-2.

SI DATI NUMERI AD ALIQUEM FUERIT PROPORTIO DATA, ET ILLUM
DATUM ESSE CONSEQUITUR. In multiplici proportione usque facile. In aliis autem facile quidem, si consequens datur, quoniam ad ipsum referuntur partes quas datas esse haut absurdum.
5 Ipsum ergo multiplicabitur si necesse fuerit, et partes quas
oporteat adiungantur et habebitur antecedens. Si vero antecedens detur idem dividetur per denominationem, et exibit consequens. Vel aliter. Sumetur numerus qui huius partes habet,
poneturque consequens, et invenietur antecedens in illa propor-
10 tione, et sic praemissa operare. Verbi gratia: Sit numerus qui
cum tanto et iterum tanto atque dimidio et dimidii dimidio faciat
c. Dividatur ergo c per tria et dimidium et quartam, et exibunt
xxvi et duae tertiae, et hoc est consequens. Item est numerus

cuius quarta et sexagesima sit xxvi et duae tertiae. Sumatur
15 numerus qui habet quartam et sexagesimam, et ipse est lx, cuius
quarta et sexagesima est xvi. Ducatur ergo lx in xxvi et duas
tertias, et fient mdc. Hic dividatur per xvi, et exibit c qui
est consequens.

II-3.

SI PRIMI AD SECUNDUM FUERIT PROPORTIO DATA, ET SECUNDI
AD PRIMUM PROPORTIO DATA ERIT. Dividatur enim unum per denominationem proportionis primi ad secundum et quod exierit,
erit denominatio secundi ad primum. Verbi gratia: Primum con-
5 tineat secundum bis et eius duas tertias. Per duo ergo et duas
tertias dividatur unum, et exibunt tres octavae. Erit ergo
secundus tres octavae primi.

II-4.

SI TOTIUS AD DETRACTUM PROPORTIO DATA, RESIDUI AD DETRAC-
TUM PROPORTIO DATA. QUOD SI RESIDUI AD DETRACTUM DATA FUERIT
PROPORTIO, ET TOTIUS AD DETRACTUM SIMILITER DATA ERIT. Hoc
facile est. Si enim a proportione totius ad detractum tolla-
5 tur unum, remanebit proportio residui ad detractum. Si item
proportioni, quae est residui ad detractum, addatur unum. Fiet
proportio totius ad detractum. Verbi gratia: x continet tria
ter et eorum tertiam. Itaque vii continet tria bis et eorum
tertiam. Converso modo vii continet tria bis et tertiam.
10 Ergo x continet tria ter et insuper tertiam.

II-5.

SI TOTIUS AD DETRACTUM FUERIT PROPORTIO DATA, ET TOTIUS AD
RESIDUUM ERIT PROPORTIO DATA. Si enim totius ad detractum
fuerit proportio data, et residui ad detractum erit data.
Quare detracti ad residuum, ergo et totius ad residuum. Ver-
5 bi gratia: x continet vi et eius duas tertias. Ergo iiii
est duae tertiae. Dividatur ergo unum per duas tertias et
exibit unum et medietas. Quare vi continet iiii semel et
dimidium. Ergo x bis et dimidium.

II-6.

SI NUMERUS DATUS DIVIDATUR IN DUO, QUORUM PROPORTIO
FUERIT DATA, UTRUMQUE EORUM DATUM ERIT. Si enim proportio
unius ad reliquum data fuerit, et totius ad idem data erit

proportio. Cum ergo totum sit datum, erit et illud datum et
5 ob hoc reliquum. Verbi gratia : Dividatur x in duo, quorum
unum quadruplum alteri. Itaque x erit ei quintuplum, et ipsum
est duo.

II-7.

SI PRIMUM AD SECUNDUM DATUM, ET AD QUOD SECUNDUM HABET
PROPORTIONEM ERIT DATUM. QUOD SI AD ILLUD FUERIT DATUM, ET AD
SECUNDUM DATUM ERIT. Denominatio enim proportionis primi ad
secundum, in denominationem proportionis secundi ad tertium
5 ducatur, et fiet proportio primi ad tertium. Item proportio
secundi ad tertium dividat proportionem primi ad tertium, et
exibit proportio primi ad secundum. Verbi gratia : Primum continet secundum et eius tres septimas, et secundus tertium et
eius duas quintas. Ducatur ergo unum et tres septimae in unum
10 et duas quintas, et provenient duo, quare primum est duplum
tertie. Item duo dividantur per unum et duas quintas, et exibunt unum et tres septimae. Itaque aliis positis primum continebit secundum et eius tres septimas.

II-8.

SI QUILIBET NUMERI AD UNUM PROPORTIONEM HABUERINT DATAM,
ET TOTUM EORUM AD EUNDEM PROPORTIONEM HABEBIT DATAM. Denomina-
tiones proportionum omnium ad illum coniungantur, et compositum
erit denominatio totius ad idem. Verbi gratia : Primum contin-
5 eat quartum semel et tertiam, secundum bis et quartam, tertium
bis et dimidium, quae faciunt vi et duodecimam. Quare primum,
secundum et tertium continebunt quartum sexies et eius duodecimam.

II-9.

SI UNUS NUMERUS AD QUOTLIBET PROPORTIONEM HABUERIT DATAM,
ET AD EX ILLIS COMPOSITUM PROPORTIONEM HABEBIT DATAM. Si enim
ille ad illos, etiam ipsi ad eum proportionem habebunt datam.
Quare et compositus eorum ad eundem, ergo et ipse ad compositum.
5 Verbi gratia. Sit ut unum contineat et eius duas tertias, et
alium et eius dimidium. Dividatur itaque unum per unum et duas
tertias, et exibunt tres quintae. Et iterum per unum et dimidium, et exibunt duae tertiae. Coniunctim erunt unum et quinta
et quintadecima. Per hoc dividatur unum, et exibunt xv undevige-
10 simae. Itaque illud erit xv undevigesimae illorum coniunctorum.

II-10.

SI A DUOBUS NUMERIS DATIS DUO NUMERI DETRAHANTUR, FUERITQUE
DETRACTORUM ET RESIDUORUM PROPORTIO DATA, NON AUTEM QUAE TOTIUS
AD TOTUM, ERIT ETIAM QUODLIBET EORUM DATUM. Sint numeri dati
ab, *cd*, detracta *a*, *c*. Et quia *a* ad *c* non sicut totum ad totum,
5 non erit *a* ad *c* sicut *b* ad *d*. Sit igitur *a* ad *e* sicut *b* ad
d. Erit ergo *ab* ad *ed* sicut *b* ad *d*, et quia *ab* datum, similiter et *ed*. Quare et differentia *c* ad *e* data quae sit *g*. Cum
sit igitur *a* ad *c* et ad *e* proportio data erit ad *g*. Sed *g* est
datum, quare et *a*, singula ergo data. Verbi gratia: Sint dati
10 numeri xx et xii, et detractum xx sit duplum detracto xii, et
residuum xx sesquialterum residui xii. Sit autem xx duplum ad
quiddam et ipsum est x cuius differentia ad xii est duo. Et quia
siquidem est sesquialterum ad totum, et duplum ad detractum, est
sescuplum ad reliquum, erit residuum xx sescuplum ad duo. Ergo
15 erit xii, et detractum viii. Detractum vero xii erit iiii, et
residuum ipsius erit viii.

II-11.

SI DUO NUMERI FUERINT AD INVICEM DATI, NUMEROQUE DATO
AB ALTERO DETRACTO ALTERIQUE ALIO DATO ADDITO SI POSTMODUM PRO-
PORTIONEM HABUERINT DATAM, UT PRIUS SUMPTA DATA ERUNT. Sint dati
numeri *ab*, *c*, et ab *ab* detrahatur *a* datus numerus, et *c* ad-
5 datur *d* datus numerus, ut sit etiam *b* ad *cd* proportio data.
Sit item *e* ad *a* sicut *cd* ad *b*. Quare *cde* ad *ab* erit proportio data. Sed *c* ad *ab* proportio data, ergo *de* ad *ab* proportio data erit, atque *de* est datus, ergo et *ab* atque *c* dati
erunt. Verbi gratia: Sit maior minor sesquitertius, maiori-
10 que detracto vii et alii addito vi, fiat totum minoris duplum residuo alterius. Sit igitur numerus duplus vii, et ipse
est xiiii, qui addatur toti minoris, et compositus fiet
duplus maiori additque super minorem xx, et quia minor est
tres quartae maioris, auferantur tres quartae de duobus, et
15 remanebunt v quartae. Itaque xx continet maiorem semel et
quartam, et ipse est xvi, minor itaque erit xii.

II-12α.

SI DUOBUS NUMERIS DATI NUMERI ALTERNATIM ADDANTUR ET
DETRAHANTUR, ET POST MUTUAM ADDITIONEM ET DETRACTIONEM SINT
SEMPER AD INVICEM DATI, UTERQUE ERIT DATUS. Sint numeri *ab*

et *de*, et dati sint *a* et *d*, itemque dati *c* et *f*. Si ergo
5 *abc* fuerit datus ad *e*, etiam *def* ad *b*, erunt *ab* et *de* dati.
Quia enim *abc* est datus ad *e*, detractoque ab eo dato numero
ac et alteri dato addito qui est *df*, fit *def* ad *b* datus per
praemissam operationem. Verbi gratia: Minori detrahatur iiii,
et alii additis duobus sit totum maioris duplum residuo al-
10 terius, atque minori additis tribus et maiori demptis iiii
sit totum residuo sesquitertium. Per operationem ergo praemissae totum minoris erit xvi, et maioris residuum xii. Maius
ergo xvi, et minus xii non equales.

II-12β.

SI A DUOBUS NUMERIS AD INVICEM DATIS DUO DATI NUMERI DE-
TRAHANTUR, UT SINT RELIQUI AD SE DATI, ET TOTI NUMERI DATI ERUNT.
Sint numeri ad invicem *ab* et *cd* atque *a* et *c* sint dati, et *b*
et ⟨*d*⟩ ad se dati. Sit itaque *e* ad *a* sicut *d* ad *b*, differenti-
5 aque *e* ad *c* sit *g*. Eritque totus *ed* ad *ab* datus, quare et ad
cd. Itaque et *g* ad *cd* est data. Et quia *g* data erit, et *cd*
et *ab* datus. Verbi gratia: Sit maior duplus minori et a maiore iiii et a minore detrahantur vi, ut sit residuum quadruplum
residuo. Sitque numerus quadruplus vi, hoc est xxiv, qui adda-
10 tur residuo maioris, ut sit totus quadruplus minori et maiori
duplus. Quare differentia, quae est xx, est ei equalis et ipse
erit xx et alter x.

II-13α.

SI A DUOBUS NUMERIS DATIS DUO NUMERI AD INVICEM DATI DE-
TRAHANTUR, ET RESIDUORUM SIT DIFFERENTIA DATA, SINGULA EO-
RUM DATA ESSE NECESSE EST. Sint dati numeri *ab* et *cd*, atque
a ad *c* datus, et differentia *b* ad *d* data, quae sit *e*. Sitque
5 *f* differentia, quam addit *ab* super *cd*, et differentia *a* et *c*
sit *g*. Si igitur *b* maius *d* atque *e* maior *f* sive minor quam
eorum differentia et *a* et *c*. Quod si *b* minus *d*, tunc *e* et *f*
facient *g*, quare semper *g* datum, ergo et *a* et *c*, sicque *b* et
d. Verbi gratia: Dati sint xv et ix. Detractum a xv
10 sit triplum ad detractum de ix, et differentia residui ix ad
residuum xv sit duo quae addatur super differentiam xv ad ix,
et fient viii quae est differentia detractorum. Ideoque erunt
xii et iiii, residua iii et v.

II-13β

DATIS DUOBUS NUMERIS AD INVICEM, SI AB ALTERO NUMERUS
DATUS ET A RELIQUO NUMERUS AD IPSUM DATUS DETRAHANTUR UT SINT
RESIDUI AD SE DATI, ERUNT ET TOTI NUMERI DATI. Si enim unus ad
suum detractum datus erit et ad residuum. Quare et ad residuum
5 alterius. Et quia ad totum erit similiter et ad eius detractum
quod est datum. Quare et ipse datus atque reliquus. Verbi
gratia: Numerus ad alium sesquialter et a maiori iiii ab alio
quarta et octava detrahantur, ut sit residuus residuo duplum.
Quia igitur detractum minoris est eius tres octavae, erit resi-
10 duum eius v octavae, quod cum sit residui dimidium residui maioris. Erit residuum tamquam totum minus et eius quarta. Detractum ergo maioris hoc est iiii erit eius altera quarta. Ipsum
ergo erit xvi et alter xxiiii.

II-14α.

SI A DUOBUS NUMERIS DATIS NUMERI DEMANTUR, QUI SINT
AD SE DATI, ET EX RELIQUO IN RELIQUUM DATUS NUMERUS PROVENIET,
QUILIBET EORUM DATUS ERIT. Sint dati numeri *ab*, *de* et *a* ad *d*
datus, et quod fit ex *b* in *e* sit datum, quod sit *f*. Sit autem
5 sicut *a* ad *d* ita *abc* ad *de*. Quare *bc* ad *e* datum est. Ergo quod
fit ex *b* in *bc* sic datum ad *f*, et differentia *bc* ad *b*, quae est
c, sic data. Erit et *b* et *bc* datum, et sic omnia. Verbi gratia:
Sint dati xii et x, demptumque a xii ad demptum ex x sit sesqualterum, at ex reliquo in reliquum fiant xxxvi. Est autem xv
10 ad x sesqualterum et liiii ad xxxvi sesqualterum, qui cum con-
tineatur sub duobus numeris, quorum differentia iii, que est xv
ad xii, erit unus ix et alter vi, qui est una portio xii, et alia
vi; portionesque x erunt iiii et vi.

II-14β.

SI FUERINT DUO NUMERI AD SE DATI, ET AB ALTERO AUFERA-
TUR DATUS NUMERUS, RESIDUIQUE QUOTALIBET PARS RELIQUO ADDITA
TOTUM QUID FECERIT DATUM, UTERQUE EORUM DATUS ERIT. Sint duo
numeri *ab* et *c* et sit *a* datus, et quarta pars *b* addita *c* faciat
5 totum *cd* datum. Igitur *cd* quadruplicetur ut fiat *ef*, et *f*
quadruplum *c* atque *e* equale *b*. Equale autem *a*, qui sit *h*,
addatur *ef*. Eritque *he* equale *ab*. Et quia *ab* ad *c* est datus,
erit et *he* ad eius quadruplum hoc est *f*. Totius ergo ad *hef*

ad *he* proportio data. Cumque sit *hef* datus erit et *he* atque
10 *ab* datus. Verbi gratia: Sit unus ad alium sesquiquartus et
detracto ab eo iiii reliqui medietas ei addita faciat xi.
Dupletur ergo xi et fiant xxii cui addatur iiii et erunt
xxvi. Quia ergo duplum subsesquioctava iii est continens
alium et eius tres quintas erit xxvi illum continens bis et
15 eius tres quintas. Ille ergo erit x et alius viii.

II-15.

SI NUMERUS IN QUOTLIBET DIVIDATUR QUORUM UNUM DATUM, ET
AD RELIQUORUM SINGULA PROPORTIONEM HABUERIT DATAM, ERIT TOTUS
NUMERUS DATUS. Si enim ad singula eorum proportionem habuerit
datam, et ad compositum ex eis, quare et ad residuum, et quia
5 illud est datum, erit ipsum totum datum. Verbi gratia: Dividatur totum in iiii quorum unum est tertia, alterum quarta,
alterum quinta, et relinquentur vi et dimidium. Sed tertia,
quarta et quinta sunt xlvii sexagesimae. Erunt ergo vi et dimidium xiii sexagesimae. Sed unus numerus xxx. Ergo xxx divisus
10 est cuius portiones x, vii et dimidium, vi, et remanebunt vi et
dimidium.

II-16.

NUMERO IN QUOTLIBET DIVISO SI UNUM EORUM CUM DATO NUMERO
FECERIT NUMERUM, QUI AD TOTUM SIT DATUS, RELIQUIS AD IPSUM DA-
TAS HABENTIBUS PROPORTIONES, IPSE NUMERUS DIVISUS DATUS ERIT.
Cum enim singula eorum ad divisum proportiones habeant datas,
5 et compositus ex omnibus ad eum proportionem habebit datam.
Quare et residuum quod est datus numerus habebit ad eum proportionem datam. Quare et ipse datus erit. Verbi gratia: Dividatur numerus in tria, quorum unum sit tertia, aliud quarta, reliquum cum tenario sit eius duo tertiae. Quare omnia cum tena-
10 rio continebunt eum et eius quartam, tenarius ergo eius quarta,
et ipse erit xii.

II-17.

DIVISO NUMERO IN QUOTLIBET, SI EORUM ALIQUOD CUM NUMERO
AD TOTUM DATO FECERIT NUMERUM DATUM ET RELIQUA AD TOTUM PROPOR-
TIONES DATAS HABUERINT, IPSUM NUMERUM DIVISUM DATUM ESSE CONVE-
NIET. Cum enim ad singula reliquorum proportionem datam habeat,
5 et ad coniuncta habebit, et sic ad prius sumptum et ei additum.

Quare et ad compositum habebit proportionem datam, quod cum datum sit, et ipsum datum erit. Verbi gratia: Dividatur, ut prius,
in tria, et unum sit medietas, et unum tertia, et tertium cum
eius quarta faciat v. Et quia dimidium et tertia sunt v sextae,
10 erit reliquum sexta, quae, quoniam cum quarta facit v, erit v
eius quarta et sexta, hoc est v duodecimae, quare ipse est xii.

    II-18.

SI NUMERUS DATUS IN QUOTLIBET DIVIDATUR, QUAE CONTINUE
PROPORTIONEM HABUERINT DATAM, ERIT QUODLIBET EORUM DATUM. Quia
enim primum ad secundum datum et secundum ad tertium, erit primum ad tertium datum, quare similiter ad secundum et tertium, et
5 ob hoc totum ad ipsum datum, et cum totum datum sit, erit etiam et
primum, et ideo secundum atque tertium. Verbi gratia: Dividatur lx in tria, quorum maius duplum secundo, secundum triplum
tertio. Quare primum sescuplum tertio, quare secundum et tertium duae tertiae primi. Totum ergo continet primum et eius
10 duas tertias. Erit ergo illud xxxvi, secundum xviii, tertium vi,
quae simul faciunt lx.

    II-19.

NUMERO DATO PER QUOTLIBET DIVISO SI SINGULIS EORUM
DATI NUMERI ADDANTUR, UT COMPOSITORUM SIT CONTINUE PROPORTIO
DATA, ERIT PRIUS SUMPTORUM DIVIDENTIUM PROPORTIO DATA SIMILI-
TER, ET IPSA DATA. Numeri dati pariter additi dato numero di-
5 viso addantur, et fiet numerus totus datus, quam constat divisum esse in numeros ad se continue datos, qui constant ex divi-
dentibus primi numeri et numeris datis. Quare compositos illos
datos esse constat, ex quibus si dati numeri auferantur, relinquentur portiones propositi numeri dati. Verbi gratia: xx di-
10 vidatur in tria, quorum uni addantur iiii, alii unum, alii v, ut
sit primum secundo sesqualterum, secundum tertio duplum. Itaque
iiii, v, unum addantur xx, et fient xxx. Quod si divisum fuerit
secundum illas portiones, exibunt xv, x et v. Auferantur ab illis
dati numeri, a xv iiii, et a x v, et a v unum, et remanebunt xi,
15 v, iiii, in quae constat divisum esse xx.

    II-20.

SI SUMANTUR QUOTLIBET NUMERI, QUORUM PRAECEDENTES CUM
DATIS NUMERIS AD SEQUENTES HABUERINT PROPORTIONES DATAS ITA, UT

ET ULTIMUS ADIUNCTO DATO NUMERO HABEAT PROPORTIONEM DATAM AD PRI-
MUM, SINGULOS EORUM DATOS ESSE DEMONSTRABITUR. Sint tres numeri
5 *a, b, c,* et dati numeri totidem *d, e, f,* sitque *ad* ad *b,* et *eb* ad
c, et *fc* ad *a* dati. Sicut igitur *ad* ad *b* ita sit *g* ad *e,* ut sit
totus *gda* ad *bc* sicut *da* ad *b,* et hoc est datus. Cum ergo sit
be ad *c* datus, *gda* etiam datus ad *c.* Sit item *h* ad *f* sicut
gda ad *c,* critque *hgda* ad *fc* datus. Sed *fc* ad *a* datus, ergo
10 *hgda* ad *a* datus est, quare et *hgd* ad *a* datus. Sed *hgd* datus.
Ergo et *a* datus, similiter *b* et *c.* Verbi gratia: Sint numeri
iii, sitque primus vi continens secundum et eius duas tertias,
secundum cum iiii continens tertium cum bis, tertius cum duobus
sit v septimae primi. Et quia vi et duae tertiae continent iiii
15 et eius duas tertias, addantur primo, ut sint ipse et xii et duae
tertiae continentes secundum cum iiii et eorum duas tertias.
Continebunt ergo tertium ter et eius tertiam. Sed et vi et duae
tertiae continent duo bis et eius tertiam, itaque primus cum xix
et tertia continent tertium cum duobus ter et eius tertiam. Sed
20 tertius cum duobus cum sit v septimae primi, primus et xix et
tertia continebunt primum bis et eius duas septimas et duas vigesimas primas. Quare xix et tertia continebunt eum semel, et
cetera. Erit ergo xiiii, et secundus xii, tertius autem viii.

   II-21.

SI NUMERUS DATUS IN DUO DIVIDATUR, QUORUM ALTERUM VEL NU-
MERUS AD IPSUM DATUS CUM NUMERO DATO, AD RELIQUUM FECERIT NUME-
RUM DATUM, UTRUMQUE DATUM ERIT. Dividatur *ab* in *a* et *b,* et sit
c datum ad *b,* atque *ac* sit numerus datus. Cuius differentia ad
5 *ab* sit *e* data, et ipsa est differentia *c* ad *b.* Et quia *c* ad *b*
datum, erit *e* ad *b* datum. Cumque sit *e* datus, erit et *b* et *a*
datus. Verbi gratia: xii dividatur in duo, quorum unum cum tertia reliqui faciat vi, cuius ad xii differentia est vi, qui etiam
est differentia tertiae ad totum suum. Quare vi duae tertiae, et
10 totum ix, residuum iii, quod cum tertia ix, hoc est cum tribus,
facit vi.

   II-22.

NUMERO DATO IN DUO DIVISO SI ALTERUM, VEL NUMERUS AD IP-
SUM DATUS, CUM NUMERO DATO FECERIT NUMERUM AD RELIQUUM DATUM, UT-
TRUMQUE EORUM DATUM ERIT. Ut prius *ab* datus parciatur in *a* et in

b, atque *a* cum *c* dato faciat *ac* numerum ad *b* datum. Addatur ergo
5 *c* ad *ab*, ut sit totus datus *abc*, qui divisus est in *b* et *ac*, quorum proportio data. Utrumque ergo datum. Verbi gratia: xii dividatur in duo, quorum alterum cum duobus faciat tres quartas. Reliqui duo iungantur xii et fient xiiii, qui divisus erit secundum
illam proportionem, quare alterum erit viii, alterum vi, a quo
10 subtractis duobus remanent iiii.

 II-23.

SI NUMERUS DATUS IN QUOTLIBET DIVIDATUR, QUORUM QUOD-
LIBET SUMPTUM SEQUENS SEMPER AD COMPOSITUM EX RELIQUIS PRO-
PORTIONEM HABEAT DATAM, QUODLIBET EORUM DATUM ERIT. Dividā-
tur datus numerus in *a*, *b*, *c*, *d*, atque *a* ad *bc*, et *b* ad *cd*,
5 et *d* ad *ab* proportionem habeat datam. Dividatur ergo *a* in
e et *f*, sitque *b* ad *e* et *c* ad *f* sicut *bc* ad *a*. Item quia *b*
ad *cd* habet proportionem datam atque *e* ad *b* erit *e* ad *cd* datum,
quod dividatur in *gh* secundum proportionem *c* ad *d*. Cum
igitur *fg* ad *c* sit datum, atque *c* ad *ad*, erit *fg* ad *da* da-
10 tum. Resolvatur ergo *g* in *kl* secundum proportionem *d* ad *a*,
eritque *a* tamquam *hkl*, cumque sit *l* ad *a* datum, erit similiter et *hk*. Sed *hk* ad *d* datum, quare *d* ad *a*, atque *h* ad *ab*.
Quare *a* et *b* et similiter ad *c*. Omnia ergo ad se et ad totum. Quare singula data. Quod si quinque fuerint, ut *a*,
15 *b*, *c*, *d*, *e*, resolvetur modo praedicto *a* in *dea*, quare in *de*,
atque eadem ratione *e* in *dc* et *d* in *cb* et *c* in *ba*, et sic *a*
in *ba*. Erit itaque *a* ad *b* datum. Verbi gratia: Ut sit exemplum in quatuor, dividatur siquidem xxxii in quatuor, quorum
primum sit septima secundi et tertii, secundum quinta tertii
20 et quarti, tertium medietas quarti et primi. Ducatur itaque
septima in quintam et erit trigesima quinta, et septima in dimidium, et fiet decima quarta, trigesima quinta in dimidium,
et fiet septuagesima. Fietque primum tamquam sua septuagesima et decima quarta, hoc est vi septuagesimae, et septua-
25 gesima et trigesima quinta et iiii decima quarta, hoc est
viii septuagesimae. Itaque lxiv septuagesimae primi sunt
viii septuagesimae quarti. Et quia viii est octava sexaginta
quatuor, erit primus octava quarti. Sed et primus septima
secundi et tertii, quare quartus continet secundum et tertium

30 et eorum septimam. Sed secundus est quinta tertii et quarti,
et quinta quarti tamquam quinta et trigesima quinta secundi
et quinta et trigesima quinta tertii. Quare secundus est tamquam quinta et trigesima quinta sui, hoc est viii trigesimae
quintae, et duae quintae et trigesima quinta et tertii, hoc
35 est eius xv trigesimae quintae. Itaque eius xxvii trigesima
quintae sicut xv xxxv tertii. Secundus ergo ad tertium sicut
xv ad xxv. Sed secundus ad tertium et quartum sit ut xv ad
lxxv. Erit igitur ad quintum ut xv ad xlviii. Sed primus
ad quartum sicut xi ad lxviii. Quare primus ad secundum ut
40 vi ad xv. Et quia vi, xv, xxvii, xlviii faciunt xcvi, qui est
triplum xxxii, erit xxxii divisus in duo, v, ix, xvi, secundum sumptas habitudines.

II-24.

SI SUMANTUR QUOTLIBET NUMERI, QUORUM QUILIBET, CUM
NUMERO DATO HABUERIT PROPORTIONEM DATAM AD COMPOSITUM EX RE-
LIQUIS OMNIBUS, SINGULOS EORUM DATOS ESSE NECESSE EST. Sint
iiii, *a, b, c, d,* datus numerus *e.* Et quia *ae* est datus ad
5 *bcd,* esto *ae* divisus in *fgh,* ut sit *f* ad *b,* et *g* ad *c,* et
h ad *d* datus sicut *ae* ad *bcd.* Sit item *k* ad *e* sicut *f* ad
b. Et quia *be* est ad *cda* datus atque *fk* ad *be,* erit *fk* ad
cda datus. Dividatur ergo secundum proportionem eorum in
l,m, n, sitque item *c* ad *a* sicut *g* ad *c.* Quare *keac* erit
10 ut *cnlg* in *h.* Sed *eb* ad *acd* est datum, atque *acd* ad *nclg*
in *h* cui est equalis *keac.* Quare *eb* ad *keac* datum erit.
Sed *keac* ad *eam* datum, atque *acd* est datum. Quare *eb* ad
eam datum, similiter *eam* ad *ec* atque *ed.* Itaque *ea* ad
triplum *e* et *bcd* datum. Sed *ea* ad *bcd* datum. Quare
15 triplum *e* ad *bcd* datum, ipsum ergo datum et *a* datum, et
sic de omnibus divisim. Verbi gratia: Datus numerus 6,
sintque iiii numeri, quorum primus et vi sit nona reliquorum, secundus et vi sit tertia trium reliquorum, tertius et vi sit tres quintae residuorum, quartus vero et
20 vi sit reliquis coniunctis equalis. Ducatur itaque tertia
in nonam, et fiet vigesima septima. Itemque tertia ducatur
in unum et nonam, et fiet tertia et vigesima septima, quae
dividantur per nonam et vigesimam septimam, et exibunt duo

et dimidium. Quare secundus et sex continet primum et vi,
25 vii et eius dividium. Itemque nona ducatur in tres quintas,
et exibit quintadecima, atque tres quintae in unum et nonam, et erunt tres quintae et quintadecima, quae dividantur per nonam et quintadecimam, et exibunt tres et tres quartae. Quare tertius et vi continent primum et vi ter et tres
30 quartas. Quia item quartus et vi sunt ut residua semel ducenda
erit nona in unum, et fiet nona. Quod si etiam unum ducatur in unum et nonam, fiet unum et nona, quae dividentur per
duas nonas, et exibunt v. Quare quartum et vi est quintuplum
primo cum vi. Itaque secundum, tertium, quartum et ter vi
35 continent primum et sex undecies et quartam. Sed illa tria
continent ea novies. Quare ter vi, hoc est xviii, continet
primum et sex bis et quartam, sunt ergo viii. Demptisque vi
remanent duo, quod est primum. Sed et secundum cum vi erit
xx, quare secundum est xiiii, tertium cum vi xxx, atque ipsum
40 xxiiii, quartum cum vi xl, quia quintuplum viii, ipsum ergo
xxxiiii.

II-25.

SI PROPONANTUR IN ORDINE QUOTLIBET NUMERI, QUORUM SIN-
GULI CUM NUMERO AD PROXIMUM SEQUENTEM DATO DATUM NUMERUM
CONSTITUANT, QUILIBET EORUM DATUS ERIT. Sit datus numerus
e et quatuor numeri a, b, c, d, atque f datus ad a, et g ad
5 b, et h ad c, et k ad d. Atque a cum g et b cum h, et c cum
k et d cum f faciat e. Sit autem g ad b, ita sit l ad h, et
sicut l ad c, ita m ad k, et sicut m ad d ita n ad a. Itaque gl et lm et mn dati erunt. Et quia est differentia ag
ad gl, ea est a ad l, erit differentia a ad l data. Simili-
10 ter differentia l ad n. Quare et differentia a ad n data.
Sed cum sit a ad n datus, erit et a et n ad differentiam
datus, uterque ergo datus. Cumque sit a datus, si detrahatur ab e, relinquetur g datus. Quare et b datus erit, et
reliqui similiter. Quod si quinque fuerint positi a, g, l,
15 m, n, t, cum sit differentia a ad l data atque l ad n, erit
et a ad n differentia data. Cumque nc datum sit, si sit a
minus n et dato minus, ipso detracto de nc dato remanebit ca
datus. Quare utrumque datum. Si vero maius, quia dato maius,

ipso additio ad *nc* fiet *ac* similiter datum, et utrumque
20 datum. Verbi gratia: Datus numerus sit cxix, sintque alii
quatuor numeri, quorum primus cum dimidio secundi faciat cxix,
secundus cum tertia tertii, tertius cum quarta quarti, quartus
cum quinta primi constituant ipsum. Ducatur itaque medietas in
tertiam, et erit sexta, et sexta in quartam et fiet vigesima
25 quarta, et vigesima quarta in quintam et proveniet centesima
vigesima. Quia igitur primus et medietas secundi sunt cxix, et
medietas secundi et sexta tertii sunt lix et dimidium, primus
excedit sextam tertii lix et dimidio, et qui vi tertii et xxiv
quarta sunt x et ix et quinque sextae, vigesima quarta quarti
30 et centesima vigesima primi sunt iiii et xxiii vigesimae quartae. Quare differentia sextae tertii et centesima vigesima
primi est xiiii et xxi vigesimae quartae. Differentia ergo
primi ad centesimae vigesimae xi est lxxiiii et ix vigesimae
quartae. Quod si multiplicetur per cxx, fiet viiidcccc et xxv.
35 Et quia primus ad illud sicut cxx ad cxix, si productus dividatur per cxix, exibit lxxv, et ipse est primus. Quo detracto
de cxix remanebunt xliiii quo duplicato fiet secundus lxxxviii.
Hoc item dempto de cxix relinquentur xxxi quo triplicato fiet
tertius xciii, et isto sublato residuus erit xxvi, quo quater
40 sumpto fiet quartus ciiii, sed et si iste auferatur de cxix,
remanebunt xv qui est quinta, lxxy qui est primus.

II-26.

POSITIS QUOTLIBET NUMERIS, SI QUILIBET EORUM CUM NUM-
ERO AD COMPOSITUM EX RELIQUIS DATO FECERIT NUMERUM DATUM, SIN-
GULI EORUM DATI ERUNT. Si datus numerus *z*, et numeri propositi
a, *b*, *c*, *d* sitque *efg* datus ad *abc*, qui cum *d* faciat *z*, atque *hkl*
5 datus ad *bcd*, qui cum *a* faciat *z*, sed et *mno* datus ad *cda*, et ipse
cum *b* faciat *z* atque *pqr* datus ad *dab* qui cum *c* faciat *z*. Sit-
quae *hkl* ad *bcd* minor proportio quam *mno* ad *cda*, et istorum
minorum quam *pqr* ad *dab*, et horum minor quam *efg* ad *abc*. Quia
igitur *k*, *l* minores sunt quam *n*, *o*, tollantur ab eis et rema-
10 nent *fc*. Eritque *ah* maius *mb*, sed si *h* est maius *b*, erit *a*
minor *m*, quoniam maior est proportio *m* ad *a* quam *h* ad *b*. Ablatis ergo *a* de *m* et *b* de *h* remaneant *xv*, eritque *v* tamquam
x, *s*, *t*. Et quia *mo* minus *pr* et *b* minor *q*, quoniam minor *h*,

erit *c* minor *n*. Et quia *pq* minus *ef* et *c* minor *g*, quia minor
15 *n*, erit *d* minor *r*. Sic ergo, quia *v* resolvitur in *xst*, quae
sunt data ad *acd*, similiter et *b* resolvetur in tria data ad
acd, atque quodlibet datum *c* in tria ad *abd*. Sed et quodlibet datum *d* in tria data ad *abc*, et sic tantum *a* non resolvitur. Atque si *h* esset minus *b*, fieretque *a* maior *m*, et tunc
20 *abc* resolvetur in reliquis et *nd*. Si ergo *a* non resolvitur,
quodlibet reducetur in datam ad *a* et datum ad unum reliquorum.
Est enim *v* tamquam *xst*. Sed sunt tamquam *lzy* data ad *abc*.
Dempto ergo *y* de *v* remanet *fe*. Eritque *fe* ut *xz* et *ti*, eritque *ti* erit ut *ne*, *pe* data ad *ab*. Dempto ergo *pe* de *fe* relin-
25 quatur *re*, eritque *re* tamquam *tipe*. Sed cum sit *b* ad *re* et
tipe datum ad *a*, erit *b* datum ad *a* et ita quodlibet eorum datum
ad *a*. Erit ergo *a* datum ad *hkl*. Quare et ad *z* compositum.
Singula igitur eorum data. Verbi gratia: Sit datus numerus
xxviii, et sint quatuor numeri, quorum primus cum triplo reli-
30 quorum faciat xxviii, secundus autem cum triplo reliquorum et
quarta, tertius cum triplo residuorum et iiii septimis, quartus cum quadruplo reliquorum xxviii constituat. Ablatis igitur primo de triplo et quarta atque tribus de tribus et quarta,
secundo autem de triplo ipsius, remanebit duplum secundi
35 quantum duplum primi et quarta atque quarta tertii et quarti.
Eadem ratione duplum tertii et quarta erit, ut duplum secundi
et iv septimae et ix vigesimae octavae primi et quarti. Et
quia duo et quarta continent quartam novies, si dividatur per
novem duo et iv septimae et cum ix vigesimae octave fiet quarta
40 tertii ut duae septimae secundi et vigesima octava primi et
quarti. Demptis quoque duobus septimis de duobus, fient unum
et quinque septimae secundi tamquam duplum primi et quarta
et vigesima octava, itemque quarta et vigesima octava quarti.
Verum quia duplum quarti et iiii septimae quarti sunt ut trip-
45 lum tertii et tres septimae primi et secundi. Et quia duo et
iiii septimae continent quartam et vigesimam octavam novies,
erit quarta et vigesima octava quarti quantum tertia tertii et
vigesima prima primi et secundi. Et quia duplum tertii et
quarta continet tertiam sexies et eius tres quartas, erit ter-
50 tia tertii ut viii vigesimae primae secundi et una vigesima

prima primi et quarti. Sublata igitur vigesima prima quarti
de quarta et vigesima octava, remanebunt xxiv octuagesimae
quartae quarti ut ix vigesimae primae secundi et duae vigesimae primae primi. Sed xxiv octuagesimae quartae continent octo
55 vigesimas octavas; quae est ablata quinquies erit quarta et vigesima octava quarti, tres septimae et tres trigesimae quintae secundi et duae vigesimae primae primi et duae centesimae
quintae eiusdem. Quia igitur secundum semel et v septimae
sunt tamquam duplum et quarta et vigesima octava primi atque
60 quarta et vigesima octava quarti. Sublatis tribus septimis et
tribus trigesimis quintis de uno et quinque septimis, remanebit unum et septima et duae trigesimae quintae secundi tamquam
duplum primi et quarta et vigesima octava et duae vigesimae
primae et duae centesimae quintae eiusdem, hoc est quingenti
65 quatuor quadringentesimae vigesimae secundi ut mille octo quadringentesimae vigesimae primi. Et quia mviii continent diiii
bis, erit secundus duplus primi. Et quia secundum semel et v
septimae, si tollatur primum bis et quarta et vigesima octavae,
quantum quarta et vigesima octava quarti, hoc est viii vigesimae
70 octavae, et quia xxxii est quadruplus viii, erit quartus quadruplus primo. Et quia duplum secundi et iiii septimae atque ix
vigesimae octavae primi et quarti sunt ut primum sexies et tres
quartae, erit duplum tertii et quarta tantumdem. Et quia vi et
tres quartae est triplum duobus et quartae, erit tertium trip-
75 lum primo. Triplum igitur secundi, tertii et quarti est ut xxviii
ad unum, quae cum faciant xxviii, erit primum unum, et ob hoc
secundum duo, tertium tria, quartum quatuor.

II-27.

OPUS AUTEM ARABUM IN PARTIBUS TANTUM CONSISTIT ESTQUE
HUIUSMODI. Exempli causa sumendi sunt iiii numeri, quorum
primus cum medietate reliquorum faciat xxxvii, secundus cum
tertia, tertius cum quarta, quartus cum quinta omnium reli-
5 quorum faciat xxxvii. Ponatur igitur numerus plures ex his
partibus habens, ut est xii, cuius medietas est vi. Tolla-
turque de xii numerus, qui cum tertia reliqui faciat vi, et
ipse est iii. Item alius, qui cum quarta residui constituat
vi, et hic est iiii. Tertius quoque, qui cum quinta reliqui

10 faciat ipsum vi, qui erit iiii et dimidium. Coniunganturque
iii, iiii, iiii et dimidium et fient xi et dimidium, quod si
venirent ad xii impossibile esset sic. Reliquum igitur ad
xii, hoc est dimidium, parciatur per tria, et eveniet sexta,
et habebimus loco primi sextam, loco secundi iii et sextam,
15 loco tertii iiii et sextam, loco quarti iiii et dimidium et
sextam. Cumque sex, qui est dimidium trium, cum sexta continet sexta trigesies septies, de quaesitis numeris, primus
erit unum, secundus xix, tertius xxv, quartus xxviii. Demonstratio. Sint quatuor numeri *a*, *b*, *c*, *d*, sitque inter
20 partes maioris denominationis *e*, quae cum *a* facit datum
numerum. Sitque *f* ut *bcd*, erit igitur *a* minimum. Esto e-
nim *g* pars *acd*, quae cum *f* coniungetur, et quia pars *cd*, quota
est *e*, est maior quam pars eorum tota est *q*, cum *b* et pars
a quota est *g*, maius quam *a* cum parte *b* quota est *e*. Dem-
25 pto ergo illis partibus de *a* et *b* erit residuum *b* maius residuo *a*. Quare et parte simili *g* detracto *ab*, quia minus
est residuum, quoque maius erit residuo *a*. Itaque *b* maior
a, alii similiter. Auferantur ergo a singulis equalia *a*,
et sint *h*, *k*, *l* relinquantur *m*, *n*, *o*. Quia igitur *b* cum
30 parte *acd* facit *ae*, et *m* cum tota parte eorundem faciet *e*,
quia *b* equale *m*, et ideo *m* cum tota parte *b* et *cd*, hoc est
cum tota parte residui sui ad *f*, faciet *e*. Similiter enim
cum parte quae adiuncta *c* faciat *ae*, faciet *e*, hoc est cum
parte *kdb*, quod est reliquum ad *f* atque *c* cum parte quae
35 cum *d* facit *ae* faciet *e*, hoc est cum parte *lbc* quod est
residuum ad *f*. Detractisque *m*, *n*, *o* de *f* remanebunt *k*, *h*,
l, quod si dividantur per numerus *bcd*, fient *hkl* equalia *a*.
Habebimus ergo *a*, et adiunctis *h*, *k*, *l* cum *m*, *n*, *o* fient
b, *c*, *d*, secundum quod in opere factum est.
     II-28.

NON DISSIMILI MODO POTEST OPUS PROCEDERE IN QUA-
LIBET PROPORTIONE ADIUNCTORUM HOC PRAEOSTENSO, QUOD AD-
IUNCTA OMNIA VEL SIMUL ERUNT MINORA VEL SIMUL MAIORA HIIS
AD QUAE ADDITA SUNT. Si enim est aliter, possibile pona-
5 tur. Sintque iiii numeri, ut supra, *a*, *b*, *c*, *d*, atque *a*
cum *efg* dato ad *bcd* fiat *z*, atque *b* cum *hkl* dato ad *cda*

faciat eundem. Sitque *efg* maior *bcd*, et *hkl* minor *cda*.
Minor est igitur *e* quam *b*, addatque *m*. Itaque *a* cum *mfg*
facit *hkl*. Et quia *a* maior *h*, erit *kl* maior *mfg*. Sed *fg*
10 maior *cd*, et *cd* maior *kl*. Quare *fg* maior *mfg*, quod est
impossibile. Reducamus igitur suprapositum exemplum, in
quo primus cum triplo reliquorum trium facit numerum datum. Sumamus itaque xxviii loco xii superius posito.
Sitque semper primum, quod cum illo additur, quod ad
15 reliqua minorem habet proportionem, si fuerit maius, ut hic
triplum. Sit igitur triplum xxviii octuaginta quatuor et
minuamus numerum a xxviii qui cum triplo et quarta residui
faciet lxxxiiii, et ipse est tres et nona. Itemque alium,
qui cum triplo et iiii septimis residui faciat lxxxiiii, et
20 hic est vi et duae nonae. Tertium autem qui cum quadruplo
residui faciat lxxxiiii, et ipse est ix et tres nonae. Quae
simul iuncta faciunt xviii et sex nonas, super quae xxviii
addit ix et tres nonas, cuius tertia est tres et nona, et
hoc erit pro primo. Ipsum etiam addatur singulis aliorum,
25 et fiet loco secundi vi et duae nonae, loco tertii ix et tres
nonae, loco quarti xii et quatuor nonae, inter quae est proportio superius inventa.

EXPLICIT LIBER SECUNDUS XXVIII PROPORTIONES CONTINENS.

Incipit ⟨liber⟩ tertius

III-1.

TRIUM NUMERORUM CONTINUE PROPORTIONALIUM, SI DUO EX-
TREMI DATI FUERINT, ET MEDIUS DATUS ERIT. Extremus in extre-
mum ducatur, et tantum erit, quantum medius in se ductus. Illius ergo extrahatur radix, et habebitur medius. Verbi gratia:
5 ix et iii extremi sint. Ducaturque unus in alium et fient
xxxvi cuius radix est vi, et ipse est medius in continua proportionalitate inter ix et iiii.

III-2.

SI TRIUM NUMERORUM CONTINUAE PROPORTIONALITATIS ME-
DIUS CUM ALTERO EXTREMORUM DATUS FUERIT, ET RELIQUUS DATUS
ERIT. Si enim medius in se ducatur, et productum per alterum
extremorum datum dividatur, exibit reliquus. Verbi gratia: Sit
5 iiii alter extremorum et vi medius. Ducatur ergo vi in se, et
fient xxxvi, qui dividantur per iiii, et exibunt ix, et ipse est
tertius in continua proportionalitate post iiii et vi.

III-3.

SI TRIUM NUMERORUM CONTINUE PROPORTIONALIUM PRIMI
AD SECUNDUM FUERIT PROPORTIO DATA, ET PRIMI AD TERTIUM DATA
ERIT. Proportio siquidem primi ad secundum in proportionem
secundi ad tertium facit proportionem primi ad tertium, in se
5 ergo ducta facit eandem. Cum ergo ipsa nota sit in se ducta,
faciet extremorum proportionem datam. Verbi gratia: Proportio primi ad secundum sit sesquitertia. Ducatur ergo unum et
tertia in se, et fient unum et duae tertiae et nona: primum
ergo continebit tertium et duas tertias et nonam ipsius.

III-4.

TRIUM NUMERORUM CONTINUAE PROPORTIONALITATIS, SI PRIMI
AD TERTIUM FUERIT PROPORTIO DATA, ET PRIMI AD SECUNDUM PRO-
PORTIO ERIT DATA. Quia enim proportio primi ad secundum in
se ducta facit proportionem primi ad tertium, si proportio-
5 nis primi ad tertium radix extrahatur, habebitur proportio
primi ad secundam data. Verbi gratia: Primus contineat
tertium bis et eius quartam, hoc est ix quartas, cuius radix est tres medietates. Quare primus continet secundum
semel et medietatem.

III-5.

SI TRIUM NUMERORUM CONTINUE PROPORTIONALIUM MEDIUS DA-
TUS FUERIT, ET COMPOSITUS EX RELIQUIS, SINGULI EORUM DATI ERUNT.
Sit a ad b sicut b ad c, sitque b datus et ac datus. Et quia
quod fit ex b in se tantum est quantum ex a in c erit, quod
5 ex a in c producitur, datum, quare et utrumque datum. Verbi
gratia: Sit xii medius, et compositus ex terminis sit xxvi
qui in se ductus faciet dclxxvi. Unus autem ductus in alium
faciat cxliiii, quo quater detracto de dclxxvi remanebit c
cuius radix est x, et ipse est differentia extremorum. Erunt
10 ergo viii et xviii.

III-6.

TRIUM NUMERORUM PROPORTIONALIUM SI COMPOSITUS EX PRIMO
ET TERTIO DATUS FUERIT AD MEDIUM, UTERQUE AD ILLUM DATUS ERIT.
Sit ac ad b datus, et quia ipsius ad b est proportio composita ex proportione autem a ad b et proportione c ad b,
5 cum sit proportio ad ad b ad unum sicut unum ad proportionem, quae est c ad b, cum sit unum quod est medium datum,
et compositum ex illis proportionibus datum, erunt utraque
data. Verbi gratia: Compositum ex terminis contineat medium
bis et eius duodecimam. Itaque duo et duodecima ducantur in
10 se et fient iiii et tria et centesimae quadragesimae quartae.
Dempto ergo quod fit ex uno in se quater, hoc est iiii, remanebunt tria et centesimae quadragesimae quartae, cuius radix est vii duodecimae. Ipsum si tollatur de duobus et duodecima, relinquentur unum et duae quartae, cuius medietas

15 est tres quartae. Quare minor erit tres quartae medii, et
medius similiter maioris.

III-7.

TRIBUS NUMERIS PROPORTIONALIBUS SI ALTER EXTREMORUM
FUERIT DATUS RELIQUUSQUE CUM MEDIO FECERIT NUMERUM DATUM,
QUILIBET EORUM DATUS ERIT. Sint proportionales a, b, c,
sitque a datus, et bc faciat numerum datum. Ducaturque

5 a in bc et fiat de, ut sit d ex ductu a in b, et e ex ductu
a in c. Ideoque et ipse e fiet ex b in se. Quare fit ex b
in se et in a, qui datus est, erit datum. Ipse ergo datus.
Verbi gratia: Alter extremorum sit ix et compositus ex reliquis xxviii. Ducatur itaque ix in xxviii, et fiet cclii quod

10 quater sumptus facit mille et viii quibus addatur quadratum
ix et fient mille lxxxviiii cuius radix est xxxiii. Sublato
ix remanent xxiiii, cuius dimidium est xii et ipse est medius trium, tertiusque erit xvi.

III-8.

SI ALTERUM EXTREMORUM CUM MEDIO AD RELIQUUM EX-
TREMORUM DATUM FUERIT, UTRUMQUE AD MEDIUM ERIT. Ut
sit ab ad c proportio data, atque ipsa constat ex proportione a ad c et b ad c. Sed proportio a ad c ad proportio-

5 nem b ad c, sicut proportio b ad c ad unum, per praemissam
ergo utraque earum data erit. Verbi gratia: Sit alterum
extremorum cum medio sesculum ad tertium. Itaque sex quater sumptum facit xxiiii, cui addito uno fient xxv, cuius
radix v, de quo dematur unum, et reliqui dimidium erunt duo,

10 quare medium minori et maius medio duplum erit.

III-9.

SI DUPLUS MEDII CUM UNO EXTREMORUM DATUM NUMERUM FE-
CERIT RELIQUO EXTREMORUM EXISTENTE DATO, SINGULI IPSORUM DATI
ERUNT. Ut si a cum duplo b fecerit numerum datum, atque c
datus fuerit. Ducatur ergo c in se et fiat d, et in a et fiet

5 e, et in b bis et fiant fg. Eritque totus $defg$ datus. Sed et
quia e, quantum quod ex b in se, erit $defg$, quod fit ex bc in
se. Extracta ergo radice habebimus bc datum. Et quia c datus, et b atque a. Verbi gratia: Alter extremorum sit duo,

The Critical Edition

duplumque medii cum reliquo faciat xvi. Ducatur ergo duo in
10 se et xvi, et fient xxxvi, cuius radix est vi, demptoque bi-
nario remanent iiii, et ipse medius, tertius viii.

III-10.

TRIBUS NUMERIS PROPORTIONALITER SUMPTIS SI COMPOSI-
TUS EX OMNIBUS DATUS FUERIT, EXTREMORUMQUE PROPORTIO DATA,
QUILIBET EORUM DATUS ERIT. Si enim proportio fuerit data, et
extremi ad medium et medii ad tertium erit proportio data.
5 Compositus ergo secundum hoc proportionaliter dividatur, et
habebimus illos tres. Verbi gratia: Compositus ex tribus sit
xix, et extremorum alter alterum contineat bis et quartam. Duorum ergo et quarta extrahatur radix, et erit unum et dimidium.
Dividatur igitur xix per tria, ut primus secundo sit sesqualterum et secundus tertio, et fient iiii, vi, viiii.

III-11.

SI COMPOSITUS EX TRIBUS NUMERIS PROPORTIONALIBUS
FUERIT DATUS ATQUE EXTREMÓRUM DIFFERENTIA DATA, IPSI ETIAM
DATI ERUNT. Data autem differentia auferatur et item addatur composito et provenient data duplum minoris termini cum
5 medio. Itemque duplum maioris cum medio. Unumque in alterum
ducatur. Et quia, quod ex duplo minoris in duplum maioris ducitur est quantum quod ex medio in se quater, erit quod producitur quantum quod ex medio in se ter et in compositum bis datum. Quare quod fit ex medio in se et in duas tertias compositi
10 datum erit, cumque tertia sic data erit, medium datum, et sic
extrema. Verbi gratia: xxxviii differentia extremorum x. Quo
detracto de xxxviii et tunc additio fient xxviii et xlviii,
unoque in alterum ducto fient mcccxliiii cuius tertia ccccxlviii.
Hic quadruplicetur, et erunt mdccxcii, cui addatur quadratum
15 duarum tertiarum xxxviii, hoc est xxv et tertia, et fient
īīccccxxxiii et duae tertiae et nona, cuius radix est xlix
et tertia; de quo ablatis xxv et tertia reliqui medietas
est xii, et ipse est medius. Compositusque ex reliquis
xxvi, quare unus viii et alter xviii.

III-12.

TRIBUS NUMERIS PROPORTIONALITER SUMPTIS SI COMPOSITUS
EX DUOBUS EXTREMIS, ITEMQUE COMPOSITUS EX MINIMO EXTREMORUM

ET MEDIO DATI FUERINT, OMNES EOS DATOS ESSE CONVENIET. Sint
tres numeri proportionales *a*, *b*, *c*, maximus *a*, sintque dati
5 *ac* et *bc*, medietasque differentiae *a* ad *c* sit *d*. Manifestum-
que quod *cd* est medietas *ac*, quadratum item *cd* est ut quadratum *b* et quadratum *d*, quam quadratum *cd* est ut quadratum *d* et
quod fit ex *d* in *c* et *cd* in *c*. Sed *cd* cum *d* facit *a*, atque
a in *c* ut *b* in se. Et quia *bc* notum et *cd* datum, erit dif-
10 ferentia eorum data, quae est differentia *d* et *b*. Quare cum
quadrata eorum data compositus ex eis et uterque datus erit.
Cumque sic *b* datus, et *ac* erit et *a* et *c* datus. Verbi gratia:
Compositus ex maximo et minimo sit xxxiiii, et ex medio et
minimo sit xxiiii, atque medietas xxxiiii est xvii cuius quad-
15 ratum est cc et lxxxix, et ipsum constat ex quadratis mediis
et dimidiae differentiae extremorum, quorum etiam differentia
est vii. Quadrato igitur vii, hoc est xlix, sublato de
cclxxxix remanebunt ccxl qui cum aliis iuncti facient dxxix
cuius radix xxiii de quo ablato vii reliquoque dimidiato fi-
20 ent viii, et residuum de xiii erit xv, qui est medius, et
sic duo extremi provenient viiii et xxv.

III-13.

SI VERO COMPOSITUS EX DUOBUS EXTREMIS ITEMQUE EX
MAXIMO ET MEDIO DATI FUERINT, TERMINOS PROPORTIONALES DU-
PLICITER ASSIGNARI CONTINGIT. Ita si *ac* et *ab* sint dati,
possibile erit dupliciter sumi *a*, *b*, *c*. Esto enim quod *d*
5 sit medietas differentiae portionum *ac*, quae sit *e* maior
et *f* minor. Semper enim *ac* maior duplo *b*. Dico ergo quod
d proportionalis erit inter *e* et *f*, atque *cd* quantum *ab*.
Quia enim *fb* est medietas *ac*, erit quadratum eius tantum
quadratum *b* et *d*. Remanet ergo quadratum *d* quantum quod
10 fit ex *b* in *f* et *fb* in *f*. Et quia *fb* et item *b* sunt ut
e erit quadratum *d* quod fit ex *e* in *f*. Quare *d* inter
e et *f*. Sed et *ed* constat ex dimidio et *b* et *d*, et *ab*.
Similiter equalia ergo sunt data, cum erunt *d* et *b*. Cum
enim *ab* datum atque dimidium, datum erit *db*. Sed et
15 quadrata eorum data, utrumque ergo datum. Ob hoc etiam
et *e* et *f*. Itaque *a* et *c* data erunt. Verbi gratia: Compositum ex maximo et medio sit xxviii et ex maximo et

minimo xxv. Dimidium autem xxv est xii et dimidium, cuius
quadratum est clvi et quarta. Eiusque differentia ad
20 xxvii xv et dimidium, huius quadratum est ccxl et quarta
de quo tollatur clvi et quarta et relinquentur lxxxiiii
cuius ad clvi et quartam differentia est lxxii et quarta,
cuius radix vii et dimidium quo dempto de xv et dimidio et
reliquo mediato exibunt tria et dimidium, quod cum xii facit
ipsum. Igitur tam xii quam tria et dimidium potest esse medium. Et si fuerit xii, erunt extrema xvi et ix; si tria et
dimidium, erunt extrema xxiiii et dimidium maius et dimidium
tantum erit minimum.

III-14.

SI FUERINT QUATUOR NUMERI PROPORTIONALES, FUERINTQUE
PRIMUS ET QUARTUS DATI ATQUE COMPOSITIUS EX SECUNDO ET TERTIO,
OMNES QUOQUE DATI ERUNT. Quia enim primus et quartus dati, et
quod fit ex primo in quartum quod ex tertio in secun-
5 dum, erit quod ex tertio in secundum producitur, datum, et cum
compositus ex ipsis datus sit, utrumque eorum datus erit. Verbi
gratia: Primus xv, quartus vi, compositus ex secundo et tertio
xix. Ducatur ergo xv in vi, et erunt xc. Sed et quadratum xix
est ccclxi de quo quater tollatur xc et remanebit unum, cuius
10 radix est unum, et ipsum est differentia tertii et secundi. Quare
ipsi erunt x et ix. Sed non est distinctio, quod sit tertium,
quod secundum.

III-15.

PRIMO AUTEM ET QUARTO DATIS SI DIFFERENTIA SECUNDI ET
TERTII DATA FUERIT, UTERQUE EORUM DATUS ERIT. Eadem enim de
causa qua et prius, quod fit ex secundo in tertium datum erit.
Cum ergo sit eorum differentia data consequitur eos datos esse.
5 Verbi gratia: Primus xii, quartus tres. Differentia secundi
et tertii quinque. Itaque ex ductu xii in tria fiunt xxxvi
quod quater sumptum cum quadrato v faciet clxix cuius radix est
xiii, de quo dempto v reliqui medietas erit iiii, qui est unus
et reliquus ix. Sed erit indistincte quis tertius, quis se-
10 cundus.

III-16.

SI ITEM PRIMUS ET QUARTUS DATI FUERINT, ET PROPORTIO
SECUNDI AD TERTIUM DATA, QUILIBET EORUM DATUS ERIT. Si enim

dati sunt primus et quartus, erit eorum proportio data, quae
constat ex proportione primi ad tertium et tertii ad secundum
5 et secundi ad quartum. Sed cum proportio tertii ad secundum
data sit divisa per ipsum proportionem primi ad quartum, data
erit et composita ex proportione primi ad tertium et secundi
ad quartum. Totius ergo radix extrahatur et habebitur proportio primi ad tertium. Quare tertium datum. Sed et proportio
10 secundi ad quartum, et ab hoc secundum datum. Verbi gratia:
Primum sit xviii, quartum duo, secundum quadruplum tertio.
Sed xviii continet duo novies. Itaque novem dividantur
per quartam et exibunt xxxvi cuius radix extrahatur,
et erit vi. Primum ergo continebit sexies tertium. Erit ergo
15 tertius tres, et secundus sexies duo, et ipse secundum hoc
erit xii.

III-17.

SI FUERINT QUATUOR NUMERI PROPORTIONALES, PRIMUSQUE ET
QUARTUS DATI FUERINT, COMPOSITUSQUE EX PRIMO ET SECUNDO AD TER-
TIUM DATUS, SINGULOS EORUM DATOS ESSE CONVENIET. Sint proportionales numeri *a, b, c, d.* Datique sint *a* et *d,* et *ab* ad *c*
5 datus. Et quia proportio *ab* ad *a* est ut proportio *b* ad *a* et
unum erit ut proportio *a* ad *b* ducta in proportionem *b* ad *a* et
in unum faciat proportionem *ab* ad *c.* Sed proportio *a* ad ⟨*c*⟩
ducta in proportionem *c* ad *d* facit proportionem *a* ad *d.* Sicut
ergo proportio *a* ad *d* ad proportionem *ab* ad *c* ita proportio *c*
10 ad *d* ad proportionem *b* ad *a* et ad unum. Sed qui proportio *c* ad
d ad unum sicut unum ad proportionem *b* ad *a,* utrumque ad medium
datum esse consequitur. Quare utraque data, et sic *b* et *c* data
erunt. Verbi gratia: Primum sit xvi, quartum tria, atque primus et secundus quadruplum sint tertio. Cumque sit xvi continens
15 iii quinquies et eius tertiam, v et tertia continebunt iiii
et eorum quartum et duodecimam. Et iiii erit tres quartae quin-
arii et unius tertiae. Itaque tres quartae quater sunt tria,
quibus addatur unum, et fient iiii, cuius radix est duo, de quo
subtracto uno et reliquo mediato proveniet medietas unius. Se-
20 cundus ergo medietas xvi et est viii. Tertius duplus tribus, et
est vi. Aliter. Sumatur quarta xvi quae est iiii sicut tertius
est primi et secundi, et ducatur iii in iiii et fient xii cuius
quadruplum addito quadrato iiii faciet lxiiii cuius radix viii

de quo demptis iiii et reliquo mediato fient duo. Quae cum
25 quatuor facient vi, et ipse est tertius, secundus viii.

III-18.

QUATUOR NUMERIS PROPORTIONALIBUS SI COMPOSITUS EX
PRIMO ET SECUNDO, ITEMQUE EX TERTIO ET QUARTO DATI FUERINT,
PRIMUSQUE AD QUARTAM DATUS, SINGULOS EORUM DATOS ESSE NE-
CESSE EST. Cum enim compositi dati sunt et proportio eorum
5 data; sed quae proportio compositi ex primo et secundo ad
compositum ex tertio et quarto, ea primi ad tertium, ergo
data, quaecumque primi ad quartum data, erit primi ad compositum ex tertio et quarto data. Datum ergo primum, sicque
tertium, sicque secundum et quartum. Verbi gratia: Compo-
10 situm ex primo et secundo xxv et ex tertio et quarto x, sed
et quartus sit quatuor quintaedecimae primi; cumque sit x
duae quintae xxv, erunt quartus et tertius decem vigesimae
quintae primi, cumque tertius, et quartus sit x, erit primus xv, secundus x, tertius vi, quartus iiii.

III-19.

SI VERO COMPOSITUS EX PRIMO ET QUARTO ATQUE EX SECUNDO
ET TERTIO DATI FUERINT, PROPORTIOQUE PRIMI AD TERTIUM DATA,
QUILIBET EORUM DATUS ERIT. Ut sit *ad* atque *bc* dati itemque
proportio *a* ad *c* data. Erit ergo proportio *ab* ad *cd* data.
5 Cumque totus *abcd* datus erunt *ab* et *cd* dati. Differentia ergo
b ad *d* data atque differentia *a* ad *c*. Sed quae proportio differentiae *a* ad *c* ad differentiam *b* ad *d*, ea est proportio *ac* ad *db*,
toto ergo *abcd* dato. Dati erunt *ac* et *bd*. Cumque differentiae
a ad *c* et *b* ad *d* data sint, eos omnes datos esse consequitur.
10 Verbi gratia: Sit primus cum quarto xvi, secundum cum tertio
xiiii, atque primus sesquialter tertio. Iuncto igitur uno cum
uno et dimidio erit compositus ex omnibus, hoc est xxx, atque
compositum ex tertio et quarto duplum sesquialter. Ipse ergo
erit xii. Sed quartus cum primo erat xvi, ergo primus superat
15 tertium iiii, ergo quartum est dimidium tertii. Ipse ergo erit
viii, et primus xii, secundus vi, quartus iiii.

III-20.

SI FUERINT QUATUOR NUMERI PROPORTIONALES, TOTOQUE EX
OMNIBUS COMPOSITO DATO FUERINT DIFFERENTIAE PRIMI AD SECUN-

DUM ET TERTII AD QUARTUM DATAE, OMNES EOS DATOS ESSE DEMON-
STRABITUR. Si enim differentiae primi ad secundum et tertii
5 ad quartum datae fuerint, erit differentia primi et tertii
ad secundum et quartum data. Quare cum compositus ex omnibus datus sit, uterque eorum datus erit. Sed unius ad alium
proportio ita primi ad secundum et tertii ad quartum, primus
ergo ad secundum et tertius ad quartum est datus. Primus
10 igitur et tertius ad differentias suas ad illos dati erunt.
Cumque sint differentiae datae, et ipsi erunt dati, et reliqui. Verbi gratia: Compositus ex omnibus sit xxxv, et differentia primi ad secundum v et tertii ad quartum duo. Primi
ergo et tertii differentia ad secundum et quartum erit vii,
15 quo subtracto de xxxv residui medietas erit xiiii, qui componitur ex secundo et quarto. Compositusque ex primo et tertio xxi, qui cum sit triplus ad vii, quae est differentia ipsius
ad xiiii, erit primus triplus ad v et tertius ad duo, quae sunt
differentiae ipsorum ad secundum et quartum. Primus ergo xv,
20 secundus x, tertius vi, quartus iiii.
  III-21.

QUATUOR NUMERIS PROPORTIONALITER DISPOSITIS ET COMPO-
SITO EX OMNIBUS DATO, SI DIFFERENTIAE PRIMI AD QUARTUM ET SE-
CUNDI AD TERTIUM DATAE FUERINT, SINGULOS EORUM DATOS ESSE CON-
SEQUITUR. Composito ex a, b, c, d dato sit e differentia a ad
5 d, h differentia b ad c data. Positio quod sit a maximus et b
maior c. Quia igitur differentia a ad b et b ad c et c ad d,
si de e tollatur h, remanebit differentia a ad b cum differentia c ad d faciens quiddam datum quod erit differentia ac ad
bd data. Cumque totus $abcd$ sit datus, erunt ac et bd dati.
10 Quia igitur differentia a ad c constat ex differentia a ad b
et h. Itemque differentia b ad d ex h et differentia c ad d
et haec quator differentiae sunt ut e et h. Erit e cum h differentia ab ad cd data. Quare et ab et cd data. Sed quae proportio ab ad cd, ea est a ad c et b ad d. Quare haec data,
15 cumque sint ac et bd dati, erunt a et c, et similiter b et d
dati. Verbi gratia: Sit compositus ex omnibus xlv, differentiaque primi ad quartum vii et secundi ad tertium duo. Depo-
sitis ergo duobus de vii remanent v, quibus detractis de xlv

reliqui medietas erit xx, et ipse componitur ex secundo et
20 quarto, primusque et tertius erunt xxv. Item iuncto vii cum
duobus faciunt ix, quibus demptis de xlv residui dimidium erit
xviii, qui constat ex tertio et quarto, et xxvii ex primo et
secundo. Et quia xxv addit super xx eius quartam, primus continebit secundum et eius quartam. Itaque xxvii continet secun-
25 dum bis et eius quartam. Ipse ergo erit xii, primusque xv,
sicque tertius x, et quartus viii.

III-22.

SI TRES NUMERI PROPORTIONALES TRIBUS ALIIS CONTINUE PRO-
PORTIONALIBUS COMPARENTUR, PRIMIQUE AD PRIMUM, ATQUE TERTII AD
TERTIUM FUERINT PROPORTIO DATA, MEDIUS QUOQUE AD MEDIUM DATUS
ERIT. Ut si a ad b sicut b ad c. Itemque d ad e sicut e ad f,
5 sitque proportio a ad d et c ad f datae, erit b ad e proportio
data. Continuentur enim proportio a ad d et proportio c ad f,
et composite extrahatur radix, et ipsa erit proportio b ad e.
Verbi gratia: Primus contineat primum et eius octavam, tertius
sit duplus tertio. Ducantur ergo duo in unum et octavam, et
10 fient duo et duo octavae, quod erit denominatio proportionis
compositae, si continuentur. Cuius extrahatur radix et provenient xii octavae, hoc est unum et dimidium. Itaque medium
continet medium semel et eius medietatem. Proportio enim ex
proportionibus extremorum continuata est tamquam proportio me-
15 diorum duplicata.

III-23.

SI QUOTLIBET NUMERI CONTINUE PROPORTIONALES TOTIDEM
ALIIS CONTINUE PROPORTIONALIBUS COMPARENTUR, FUERINTQUE PRIMI
AD PRIMUM, SECUNDI AD SECUNDUM PROPORTIONES DATAE, RELIQUO-
RUM AD RELIQUOS PER ORDINEM PROPORTIONES DATAS ESSE CONVENIET.
5 Quae enim differentia proportionis primi ad primum ad proportionem secundi ad secundum, ea erit proportionis primi ad secundum ad proportionem primi ad secundum, ea etiam proportionis secundi ad tertium ad proportionem secundi ad tertium, et
ita per ordinem. Sed quae differentia proportionis secundi ad
10 tertium ad proportionem secundi ad tertium, ea proportionis
secundi ad secundum ad proportionem tertii ad tertium. Quare
continue, quae differentia proportionis primi ad primum ad pro-

portionem secundi ad secundum, ea proportionis secundi ad secundum ad proportionem tertii ad tertium similiter in addendo
15 et diminuendo, et ita ad extremos. Illa ergo differentia continue dempta relinquetur reliquorum ad invicem proportio.
Verbi gratia: Quatuor comparantur ad iiii. Primum continet
primum ad eius tertiam, secundus est secundo equalis. Itaque
per unum, a quo denominatur equalitas, dividatur unum et ter-
20 tia, et exibunt unum et tertia, et per unum et teriam dividatur unum, et exibunt tres quartae. Tertius ergo tertii erit
tres quartae, atque tres quartae dividantur per unum et tertiam, et exibunt ix sextae decimae, quartus ergo quarti erit
ix sextaedecimae.

EXPLICIT LIBER TERCIUS XXIII PROPORTIONES CONTINENS.

Incipit quartus

IV-1.

SI DUO NUMERI PER ALIOS DUOS DIVIDANTUR, ET ISTORUM ET
ILLORUM FUERINT PROPORTIONES DATAE, DIVIDENTIA QUOQUE PROPOR-
TIONEM AD INVICEM HABEBUNT DATAM. Ut si a et b per c et d
dividantur, et exeant e et f, fuerintque a ad b et c datum ad
5 d. Dividatur proportio a ad b per proportionem c ad d, et
exibit proportio e ad f, quoniam proportio a ad b continuatur
ex proportione c ad d et proportione e ad f. Verbi gratia:
Divisoris ad divisorem sit proportio dupla et inter divisos sit
tripla proportio. Dividantur ergo tria per duo, et exibunt
10 unum et dimidium, quare inter dividentia erit proportio ses-quialtera.

IV-2.

QUOD SI INTER DIVISORES ET DIVIDENTIA FUERINT DATAE PRO-
PORTIONES, ET NUMERI DIVISI ERUNT AD INVICEM DATI. Ducatur siquidem altera in alteram et producetur illorum simili de causa.
Verbi gratia: Dividens dividenti sit sesquialterum et divisor
5 divisori sesquitertius. Itaque unum et dimidium in unum et tertiam ducantur, et fient duo. Quare numerus divisus alii duplus erit.

IV-3.

SI NUMERUS DATUS PER DUOS NUMEROS DIVIDATUR, QUORUM DIF-
FERENTIA ATQUE DIVIDENTIUM DIFFERENTIA DATAE FUERINT, IPSI ETIAM
DATAE ERUNT. Sit numerus datus a, qui dividatur per b et c,
quorum differentia d data, et exeant e et f, quorum differentia
5 g data. Sitque sicut b ad d ita h ad c, sed sicut b ad d ita f
ad g, quare f ad g sicut h ad c. Sed f in c facit a, ergo et h
in g facit a. Item b in c faciat l, igitur et d in h faciet l.
Itaque a ad l sicut g ad d. Quare a ducatur in d et productum

dividatur per *g* et exibit *l* datum. Quare cum differentia *b* ad
10 *c* sit data, erit *b* et *c* data, et ob hoc *e* et *f*. Verbi gratia:
xxiiiii dividatur per duos numeros, quorum differentia sit unum,
et exeat duo numeri, quorum differentia duo. Ducatur ergo unum
in xxiiii et erunt xxiiii qui dividatur per ii et fient xii,
quorum quadruplo addatur quadratum unius, fientque xlix cuius
15 radix vii. De quo sublato uno et reliquo mediato provenient
tria, et ipse erit minor divisorum et maior quatuor, dividentia
viii et vi.

IV-4.

SI VERO DIVIDENTIUM DIFFERENTIA DATA FUERIT, COMPOSI-
TUSQUE EX DIVISORIBUS DATUS, QUILIBET EORUM DATUS ERIT. Dispositio superior remaneat, praeter quod *bc* datus est et non *d*,
sed etiam proportio *a* ad *g*, quam *l* ad *d*. Si igitur *l* per *d* di-
5 vidatur proveniet quiddam datum, cumque *d* sit differentia *b* et
c qui faciunt unum datum, et *l* fiat ex *c* in *b*, erit *c* et *b*
datum, et sic *e* et *f* data erunt. Verbi gratia: xx dividantur
per duos numeros ex quibus componitur vii, et proveniant duo
numeri quorum differentia est vi. Dividatur ergo xx per vi,
10 et exibunt tria et tertia, cuius quadruplum est xiii et tertia
cuius quadratum si addatur quadruplo quadrati vii, fient
ccclxxiii et vii nonae, cuius radix est xix et tertia. De quo
subtracto xiii et tertia reliqui medietas est iii, quo subtracto de vii reliqui dimidium est duo, qui unus est diviso-
15 rum, reliquus v. Dividentia quoque x et iiii.

IV-5.

SI DUO FUERINT NUMERI AD INVICEM DATI, ET UNUS IN ALIUM
DUCTUS FECERIT NUMERUM DATUM, UTERQUE EORUM DATUS ERIT. Ut si
a ad *b* datus, et unus in alium fecerit *c* datum. Esto ergo
aliquis ad *c* sicut *a* ad *b*, qui fit *d* et datus. Atque ipse fiet
5 ex *a* in se. Extracta ergo radice illius habebitur *a* datus, et
sic *b* datus erit. Verbi gratia: Unus alteri sit sesquitertius,
unusque in alterum faciat xlviii. Addatur ergo xlviii sua tertia,
et fient lxiiii, cuius radix est viii, et ipse est ille unus, et
reliquus erit vi.

IV-6.

DUOBUS NUMERIS AD SE DATIS SI QUADRATA EORUM FECERINT
NUMERUM DATUM, IPSI ETIAM DATI ERUNT. Ut si *a* ad *b* datus, et

ex *a* in se fiat *c*, et ex *b* in se fiat *d*, sitque *cd* datus. Est
autem *c* ad *d* proportio *a* ad *b* duplicata. Quare et data sic
5 ergo et *c* et *d* datum erit. Verbi gratia: Alter alteri duplus
erit, et quadrata eorum coniuncta faciunt d. Quia ergo unum
uni quadruplum erit, erit d eidem quintuplum, et ipsum erit c,
cuius radix est x, et ipse est minor. Maior autem erit xx.

    IV-7.

DATIS DUOBUS NUMERIS AD INVICEM, SI QUOD FIT EX COM-
POSITO EX IPSIS IN EORUM DIFFERENTIAM DATUM ERIT, UTERQUE
EORUM ERIT DATUS. Quod enim fit ex composito in eorum differentiam est quod addit quadratum maioris super quadratum mi-
5 noris. Cumque quadrati ad quadratum sit proportio data, et
illius ad ipsum data erit. Quare quadratum datum, ob hoc latus eius et similiter reliquum. Verbi gratia: Alterum alteri
triplum sit, et compositum ex ipsis in eorum differentiam
faciet xxxii. Cum ergo quadratum quadrato sit nonuplum, et
10 ipsum erit eidem octuplum. Quare quadratum minoris erit
iiii, et ipse erit duo, et reliquus vi.

    IV-8.

SI QUADRATUS CUM ADDITIONE RADICIS SUAE PER DATUM
NUMERUM MULTIPLICATAE DATUM NUMERUM FECERIT, IPSE ETIAM DA-
TUS ERIT. Sit quadratus *a*, radix eius *b* multiplicata per
cd, ut et *c* et *d* sit eius medietas, atque ex *b* in *cd* fiat
5 *e*, sitque *ae* datus. Quia igitur *bcd* secundum *b* multiplicatus facit *ae*, quadrato *d* adiuncto ad *ae* fiant *aef*. Eritque *aef*, quod fit ex *bc* in se. Cumque sit *aef* datus, erit et
bc datus. Subtracto igitur *c* remanebit *b* datus, et sic *a*
datus erit. Verbi gratia: Sit quadratus cuius radix si mul-
10 tiplicatur per v et productum ei addatur, fient xxxvi. Cui
addatur quadratum duorum et dimidii quae sunt dimidium v,
et fient xlii et quarta, cuius radix est vi et dimidium,
de quo ablatis duobus et dimidio remanent iiii, qui est
radix et quadratus est xvi.

    IV-9.

QUADRATUM QUI CUM ADDITIONE DATI NUMERI FACIT NUMERUM
QUEM RADIX IPSIUS PER DATUM NUMERUM MULTIPLICATA PRODUCIT,
CONTINGIT DUPLICITER ASSIGNARI. Sit enim idem quadratus *a*,

radix *b*, numerus datus additus *c*, atque *dc* datus in quem
5 *b* ductus facit *ac*, cuius medietas *d*, et ipsius quadratum *f*,
atque differentia *b* ad *d* sit *g*. Quia igitur *b* in *d* bis
facit *ac*, addunt *a* et *f* super *ac* quadratum *g*. Itaque *a*
utrobique dempto addit *f* super *c* quadratum *g*. Dempto ergo
g de *d* potest remanere *b*, et addito *g* ad *d* potest fieri *b*,
10 quare dupliciter assignabitur *a*. Verbi gratia: Sit quadratus, qui cum additione viii faciat numerum, quem radix
sua per vi multiplicata producit. Medietas ergo vi, quae
est iii, in se ducta facit ix, qui addit unum super viii,
cuius radix unitas, quae erit differentia radicis prae-
15 dicti quadrati et ternarii. Hac igitur differentia dempta
et addita tenario, habebimus duo et iiii, quorum quadrata
quatuor et xvi. Utrique igitur addantur viii, et fient xii
et xxiiii, quae fiunt ex ductu senarii in duo et quatuor,
secundum quod propositum fuerat.

IV-10.

SI EX MULTIPLICATIONE RADICIS SUAE PER DATUM NUMERUM
ADDITO DATO NUMERO FIAT QUADRATUS, IPSE ETIAM DATUS ERIT.
Sit, ut prius, *a* quadratus et *b* radix, et *cd* datus numerus
multiplicans et *e* additus. Differentia igitur *b* ad *cd* sit
5 *g*, ut sit *gcd* tamquam *b*. Et quia *b* in se facit *a*, quam
etiam producit in *cd* addito *e*, constat *e* fieri ex *b* in
g, et quia *gc* in se est quantum *b* in *g* et *d* in se, erit
etiam quantum *d* in se cum *e* quam cum data sint erit et
gc datum, quare et *g* datum atque *gcd* qui est *b*, sicque *a*.
10 Verbi gratia: Est quadratus qui fit ex additione xii super
multiplicationem radicis suae per iiii. Itaque quadrato
dimidii iiii, quod est iiii, addatur super xii et fient
xvi cuius radix est iiii, de quo dimidio iiii dempto remanebunt duo, quae addito similiter iiii faciunt vi, et
15 ipse est·radix, quadratusque xxxvi.

IV-11.

SI NUMERUS AD QUADRATUM DATUS CUM ADDITIONE NUMERI AD
RADICEM IPSIUS DATI FECERIT NUMERUM DATUM, ET QUADRATUM ET
RADICEM DATOS ESSE CONSEQUITUR. Sit *a* radix et *b* quadratus,
et *c* datus ad *a*, et *d* datus ad *b*, ut sit *cd* datus. Esto

5 autem sicut *b* ad *d* ita sit *g* ad *c*. Itague *gb* ad *cd* sicut
b ad *d*, quare *bd* datus. Est autem et *g* ad *a* datus. Ipsius-
que ad illum proportio sit *e*. Quare *a* in *e* datum facit *g*,
qui cum *b* quadrato facit numerum datum. Erit igitur et *a*
et *b* datus. Verbi gratia: Tertia radicis et quarta quad-
10 rati facient xi, igitur quadratus cum radice et tertia eius
faciet xliiii. Huic itaque addatur quadratum duarum tertiarum, quae sunt dimidium unius et tertiae, et fient xliiii
et iiii nonae, cuius radix est xx tertiae, hoc est vi et
duae tertiae. Ablatis igitur duobus tertiis remanent vi,
15 et ipse est radix, quadratus vero xxxvi.

IV-12.

SI NUMERUS AD QUADRATUM DATUS CUM NUMERO DATO FECERIT
NUMERUM DATUM AD RADICEM, ET RADIX ET QUADRATUS DATUS ERIT.
Sit, ut prius, radix *a* et *b* quadratus et *d* datus ad *b*, qui
cum *c* dato numero faciet *cd* datum ad *a*. Sicut igitur *b* ad
5 *d*, ita sit *e* ad *c*, quare *e* datus, atque *bc* erit numerus datus ad *a*. Erit ergo similiter et *a* et *b* datus. Verbi gratia: Vicesima pars quadrati cum xxv faciat triplum radicis.
Itaque quadratum cum quingentis faciet sexagintuplum radicis. Medietas igitur lx in se faciet dcccc, qui addit
10 cccc super d, cuius radix xx et ipse est differentia radicis quadrati et xxx. Addita ergo hac differentia et ea
dempta a xxx habebitur radix et x et l quorum quadrata c et
īīd. Utriusque vicesima sumatur quae sunt v et cxxv
utrique additis xxv fient hinc xxx, triplum x, illinc cl
15 triplum l, ut propositum fuerat.

IV-13.

SI NUMERO AD RADICEM DATO ADDATUR NUMERUS DATUS
UT PROVENIAT NUMERUS AD QUADRATUM DATUS, UTERQUE SECUNDUM
HOC DATUS ERIT. Dispositio sit eadem, praeter quod *d* sit
datus ad *a*, et *cd* totus datus ad *b*. Eodem autem modo sicut
5 *b* ad *cd* ita sit *ef* ad *c* et *f* ad *d*. Quare *e* datus et *f* ad
a datus, atque *ef* equalis *b*. Itaque ex hoc et *a* et *b* datus. Verbi gratia: Triplum radicis cum xii facit sesquialterum quadrato, ergo duplum radicis et viii facit quad-

ratum. Secundum operationem ergo decimae praesentis pro-
10 veniant radix iiii et quadratus xvi.

IV-14.

SI COMPOSITUS EX DUOBUS NUMERIS FUERIT AD TERTIUM DA-
TUS, QUIQUE EX ILLIS PRODUCITUR AD QUADRATUM IPSIUS DATUS,
UTERQUE IPSORUM AD EUNDEM DATUS ERIT. Ut si ad *a* sit *bc*
datus, atque ex *a* in se fiat *d*, et ex *b* in *c* fiat *e* datus
5 ad *d*. Sit autem proportio *bc* ad *a*, *fg*, composita ex proportione *b* ad *a* et *c* ad *a*, proportio autem *e* ad *d* sit *h*
et ipsa producitur ex *f* in *g*. Cum ergo *fg* datum et ex *f*
in *g* fiat datum, erit et *f* et *g* datum. Itaque et *b* et
c ad *a* datum. Verbi gratia: Sit compositus ex duobus ad
10 tertium quintuplus, et quod ex uno in alterum fit, sit
quadrato eius sextuplum. Igitur vi tollatur quater de
quadrato v, et remanebit unum cuius radix unum quod tollatur de v, et reliqui medietas erit duo. Unum ergo illi
duplum, et reliquum erit triplum.

IV-15.

SI VERO COMPOSITUS AD ILLUM DATUS, ET QUADRATA EORUM
SIMILITER AD QUADRATUM ILLIUS DATA, ILLA QUOQUE AD IPSUM DATA
ERUNT. Dispositio superior remaneat, praeter quod quadrata
b et *c* sint *e* et *t* atque ex *f* in se fiat *h*, et ex *g* in se fiat
5 *l*. Eritque *h* proportio *e* ad *d* et *l* proportio *t* ad *d*, sicque *hl* datum erit. Cumque *fg* datum, erit et *f* et *g* datum,
itaque et *b* et *c* erit datum ad *a*. Verbi gratia: Compositum
sit triplum illi, et compositum ex quadratis sit additus super quadruplum quadrati eius v nonas. Quare quatuor et v
10 nonae duplentur et fient ix et nona. De quibus tollatur quadratum trium, hoc est ix, et remanebit nona cuius radix tertia.
Qua sublata de tribus et reliquo mediato perveniet unum et
tertia. Igitur unum est illi sesquitertium et alterum addet
ipsum super duas tertias.

IV-16.

QUOD SI DIFFERENTIA EORUM AD IPSUM FUERIT DATA, QUIQUE
EX UNO IN ALTERUM PRODUCITUR AD QUADRATUM DATUS, UTERQUE AD
EUM DATUS ERIT. Differentia *b* ad *c* sit *z* et cetera remaneant

ut in dispositione antepraemissae. Sit item differentia f ad
5 g, x, et ipsa erit proportio z ad a. Quia igitur ex f in g
fit h datum, differentia quoque eorum data, erunt ipsa data.
Quare b ad a et c ad a datum. Verbi gratia: Differentia duorum sit tertio equalis et, quod fit ex uno in alterum sit
quadrato eius sescuplum. Sumatur igitur vi quater addito
10 quadrato unius et fient xxv cuius radix est v. De quo dempto
uno reliqui medietas erit duo. Quare unus ad eum duplus
et reliquus erit triplus.

IV-17.

CUMQUE SIT DIFFERENTIA AD EUM DATA, SI QUADRATA SIMUL
AD QUADRATUM EIUS FUERINT DATA, IPSA ETIAM AD ILLUD DATA
ERUNT. Sit, ut prius, differentia b ad c, z, et differentia
f ad g, x quae erit proportio z ad a et c ut in dispositione
5 antepraemissae. Cumque hl, quod est quadratum f et g datum
sit et differentia eorum data, erunt ipsa data. Verbi gratia: Differentia sit illi dupla, quadrata vigesies contine-
ant quadratum; duplum autem xx est xl de quo sublato quadrato duorum relinquentur xxxvi cuius radix vi, de quo demp-
10 tis duobus residui medietas erit duo. Quare unum ei duplum
et reliquum quadruplum.

IV-18.

SI DUO NUMERI AD TERTIUM FUERINT DATI ATQUE QUI EX
IPSIS PROVENERIT, AD EUNDEM DATUS, OMNES EOS ESSE DATOS
CONVENIET. Ut si ad a dati sint b et c et ex b in c fiat
d datus ad a. Quadratus autem a sit e. Si igitur propor-
5 tio b ad a ducatur in proportionem c ad a, fiet proportio
d ad e. Quae si dividat proportionem d ad a, exibit proportio e ad a, quae ab ipso denominatur. Quare a datus.
Verbi gratia: Unus sit ei sesqualter et alter sesquitertius et productus ex eis contineat eum octuagiesquater. Unum
10 igitur et tertia in unum et medietatem et fient duo, per quae
dividantur lxxxiv et exibunt xlii et ipse est qui quaeritur.

IV-19.

SI QUOD FIT EX LATERE IN LATUS DATUM FUERIT, COMPOSITUM
EX QUADRATIS DATUS, QUODLIBET EORUM DATUS ERIT. Sint a et b
latera et quadrata c et d. Quodque fit ex a in b sit e datus,

cd dato. Et quia *e* medius est proportionaliter inter *c* et *d*,
5 cum sit datus et compositus ex illis datus, erit quodlibet datum. Verbi gratia: Unus in alium facit xxxv et quadrata similiter lxxiv cuius quadratum ⟨v̄⟩ ccccxxvi. Quadrato autem
xxxv quater ab eo detracto remanent dlxxvi cuius radix xxiv.
Quo dempto de lxxiv reliqui dimidium erit xxv, qui est minor
10 quadratorum, reliquus xlix, latera v et vii.

IV-20.

SI VERO PRODUCTO EX LATERIBUS DATO FUERIT QUADRATORUM
DIFFERENTIA DATA, SINGULUM EORUM DATUM ERIT. Dispositio eadem
superiori. Sitque *g* differentia *c* ad *d* datus. Et quia *e* proportionaliter inter *c* et *d*, erit *c* ad *e* sicut *e* ad *d*. Quare *c*
5 fit primus et *d* quartus dati. Et *e* secundus et tertius data eorum differentia. Quare omnes eos esse datos consequitur. Vel
aliter. Quia *e* datus sit, erit quod fit ex *c* in *d* datum, quod
est quadratum *e*. Eorum ergo differentia data erunt et ipsa data.
Verbi gratia: Ex uno in aliud fiat xv et differentia quadrato-
10 rum xvi. Itaque unum quadratum in aliud facit ccxxv. Quo quater sumpto addatur quadratum xvi et fient mclvi. Cuius radix
xxxiiii, de quo sublatis xvi et reliquo mediato ix minor quadratorum reliquus xxv, latera iii et v.

IV-21.

SI COMPOSITUS EX LATERIBUS ET PRODUCTUS EX QUADRATIS
DATI FUERINT, OMNES EOS DATOS ESSE CONSEQUITUR. Si enim producti extrahatur radix, ipsa erit quod producitur ex lateribus. Cumque ipsum datum sit, compositumque ex ipsis datum,
5 erunt et ipsa. Verbi gratia: Compositum ex lateribus ix, productum ex quadratis cccxxiiii cuius radix xviii et ipse est
productus ex lateribus. Cumque sit datus ex illis compositus
secundum praemissam operationem erunt ipsa iii et vi. Dempto
enim xviii quater de quadrato viiii remanent viiii quod est
quadratum differentiae eorundem.

IV-22.

QUOD SI DIFFERENTIA LATERUM SIMULQUE PRODUCTUS EX
QUADRATIS DATUS FUERIT, SINGULI IPSORUM DATI ERUNT. Simili
modo productus ex lateribus datus erit extracta radice producti ex quadratis. Cumque sit differentia eorum data, ipsa

5 etiam data esse patet. Verbi gratia: Productus ex quadratis
est c cuius radix x. Differentia laterum ⟨duorum⟩ tria, cuius
quadratum addatur quadruplo x et fient xlix. Huius radix vii
qui componitur ex lateribus. Sicque ipsi erunt duo et quinque.

IV-23.

SI ET LATERUM ET QUADRATORUM FUERIT DIFFERENTIA DATA,
ET HAEC ET ILLA DATA ESSE NECESSE EST. Differentia enim
quadratorum divisa per differentiam proveniet compositum
ex radicibus. Quod cum secundum hoc datum sit, et differ-
5 entia eorum data, utrumque erit datum. Verbi gratia: Dif-
erentia radicum est iii et differentia quadratorum li, quae
in tres dividatur, exibit xvii qui componitur ex radicibus.
Demptis ergo tribus et reliquo mediator fient vii et ipse
est minus laterum, maior vero x.

IV-24.

SI FUERINT DUO NUMERI QUORUM QUILIBET AD QUADRATUM
ALTERIUS SIT DATUS, UTERQUE EORUM DATUS ERIT. Sint duo numeri *a* et *b* et *a* datus ad quadratum *b*, qui sit *c*, et *b* datus
ad quadratum *a*, qui sit *d*. Sit igitur *e* proportio *c* ad *a* et
5 *g* proportio *d* ad *b*. Quia igitur *e* in *a* facit *c*, erit *e* ad *b*
sicut *b* ad *a*. Similiter, quia *g* in *b* facit *d*, erit *b* ad *a* sicut *a* ad *g*. Sunt igitur quatuor termini continue proportionales, *e*, *b*, *a*, *g*, quorum *e* et *g* dati. Si ergo proportionis
quae est *e* ad *g* radix cubica extrahatur, fiet proportio
10 *e* ad *b* data, quare *b* datus. Sicut igitur *g* ducatur in
quadratum *e* et producti extrahatur radix cubica, exibit *b*
datum. Similiterque *a* datum erit. Verbi gratia: Quadratum
unius continet reliquum quinquies et eius tertia, et illius
quadratum alterum decies octies continet. Ducantur ergo
15 xviii in quadratum v et tertiae quae est xxviii et iiii
nonae et fient dxii cuius latus cubicum est viii, et ipse
unius duorum; reliquum autem xii.

IV-25.

NUMERUM DATUM MULTIPLICITER SUMERE EST POSSIBILE EX
DUOBUS COMPOSITUM, EX QUIBUS PRODUCTUM HABEAT AD EUNDEM SECUN-
DUM QUAMLIBET PROPORTIONEM DATAM. Sit compositus *ab* et produc-

tum ex *a* in *b* sit *c*, cuius proportio ad *ab* data. Quadruplum au-
5 tem *c* sit *d* cuius etiam ad *ab* proportio data est, quae sit *e*.
Sumantur igitur quilibet tres termini proportionales, qui sint
f, *g*, *h*, maximus *h*. Differentiaque *h* ad *f* sit *k* atque sicut
h ad *k* ita sit aliquis ad *e* et ipse sit *l* datus, addatque *l*
super *e*, *m*. Quia igitur *l* ad *e* sicut *h* ad *k*, erit *l* ad *m*
10 sicut *h* ad *f*. Ergo sicut quadratus ad quadratum, quare *l* in
m facit quadratum. Facit igitur *l* in se *n*, atque in se tantum
facit quantum in *m* et in *e*. Sed in *e* facit *d*. Quare in *m*
faciet superfluum *n* ad *d* quod sit *z*. Eritque *z* quadratus
cuius radix sit *t*. Si igitur *l* dividatur in duo quorum dif-
15 ferentia *t*, ipsa erit ut *a* et *b*. Alterum enim in alterum
quater faciet *d*, addit enim quadratum differentiae. Quod si
alii tres sumerentur pro *f*, *g*, *h*, alius fieret *l*. Verbi gratia: Quod fit ex uno dividentium in reliquum sit duplum compositi. Quod ergo sit ex uno in aliud quater, erit eidem oc-
20 tuplum. Sunt igitur tres numeri proportionales, i, ii, iiii.
Estque differentia iiii ad unum tria. Sicut igitur quatuor
ad tria, ita erit x et duae tertiae ad viii. Ducatur igitur
x et duae tertiae in duo et duas tertias et fient xxviii et
iiii nonae cuius radix est v et tertia. Quod dempto de x et
25 duobus tertiis reliqui medietas est duo et duae tertiae quod
erit unum dividentium et reliquum viii, in quae dividitur x
et duae tertiae, quia viii in duo et duobus tertiis facit duplum x et duarum tertiarum ut proponebatur. Quod si termini
proportionales sumantur i, 3, 9, invenietur numerus compositus
30 ix divisus in vi et iii, ratione praemissa.

IV-26.

SI FUERINT DUO NUMERI QUORUM QUADRATA PARITER ACCEPTA
DATA FUERINT AD COMPOSITUM EX IPSIS, QUODQUE ETIAM AB IPSIS
CONTINETUR DATUM AD IDEM, UTRUMQUE EORUM DATUM ESSE CONSEQUI-
TUR. Sint huiusmodi numeri *a* et *b*, quadrata eorum *d* et *e*,
5 productum ex uno in aliud *c*. Cum sit ergo *de* ad *ab* datum, et
c similiter erit et duplum *c* ad idem datum, quod sit *f*, quare
et totus *def* datus ad ipsum. Sed *ab* in se facit *def*. Itaque
datus. Atque *c* inter *d* et *e* est proportionalis datumque *de*

ad *c*, erit et *d* ad *c* datus. Sed *d* ad *c* sicut *a* ad *b*. Quare
10 *a* ad *b* cum sit datus, erit et *a* et *b* datus. Verbi gratia:
Duo numeri sint, quorum quadrata continent compositum ex ipsis ter et eius quatuor septimas. Et quod fit ex uno in reliquum, continet eum semel et eius quinque septimas. Ipsum
ergo bis et duo quadrata continent eum septies. Hic igitur
15 erit vii, duo quadrata xxv, dividentia ergo tres et quatuor.

     IV-27.

SI AD UNUM NUMERUM DUOBUS NUMERIS DATIS UTRIQUE EORUM
DATUS NUMERUS ADDICIATUR, UT EX TOTO IN TOTUM PROVENIAT NUME-
RUS DATUS, ILLUM QUOQUE DATUM ESSE OSTENDETUR. Ad *a* sint *b*
et *c* dati, quibus addiciantur dati numeri *d* et *e*. Atque ex
5 *bd* in *ce* fiat *fghl*, ita, ut ex *b* in *c* sit *f* et ex *b* in *e* et
c in *d* fit *g*, *h*. Ex *d* vero in *e* fiat *l*. Quia igitur *b* et *c*
ad *a* et *g* et *h* ad *b* et *c*, erunt *g* et *h* dati ad *a* atque *f* datus ad quadratum *a*, qui sit *z*. Sed et *l* datus. Ergo subtracto ab *fghl* remanebit *fgh* datus, qui constat ex uno dato ad
10 *a* radicem et alio dato ad *z* quadratum. Datus ergo fit *f*, *h*, *a*.
Verbi gratia: Et tertiae et quartae radicis addatur unum et
ex toto in totum fiant xx. Tertia autem in quartam radicis
facit duodecimam quadrati et tertia in unum et quarta in unum
faciunt tertiam et quartam ⟨radicis⟩ et unum in unum red-
15 dit unum. Quo detracto a xx remanent xix quantum vii duodecimae radicis et duodecima quadrati. Ergo ccxxviii est ut quadratum et septem radices. Itaque trium et dimidii quadratum
addatur bis ccxxviii et fient ccxl et quarta. Cuius radix est
xv et dimidium. A quibus sublatis tribus et dimidio remanent
20 xii et ipse est radix.

     IV-28.

QUOD SI ALTERI DATUS NUMERUS ADDATUR, UT TOTUS IN RE-
LIQUUM DUCTUS FACIAT NUMERUM DATUM, ET SIC ETIAM IDEM DATUS
ERIT. Ut si *b* datus numerus addatur *d*, ut ex *bd* in *c* fiat *ef*
numerus datus. Ex *b* vero in *c* fiat *e* et ipse est datus ad *z*.
5 Ex *d* vero in *c* fiat *f*, et ipse erit datus ad *a*. Quare numerus
ad radicem datus cum numero ad quadratum dato facit numerum datum. Igitur *a* datus est. Verbi gratia: Medietas radicis addito uno in quartam facit x. Sed medietas in quartam facit oc-

tavam quadrati et unum in quartam reddit quartam radicis. Quare
10 x est ut quarta radicis et octava quadrati. Secundum hoc ergo
erit radix viii.

IV-29.

SI A DUOBUS NUMERIS AD UNUM DATIS DUO NUMERI DATI AU-
FERANTUR, UT EX DUCTU RELIQUI IN RELIQUUM PROVENIAT NUMERUS DA-
TUS, PROVENIET ET ILLE UNUS NUMERUS DATUS. Ad *a* numeri dati
sint *bc* et *de*. Sint autem numeri dati *c* et *e*. Atque ex *b* in
5 *d* proveniat *g* datus. Ex *bc* vero in *e* fiat *h* et ex *de* in *c*
proveniat *k*. Itaque ex *bc* in *de* fiat *l*, et ex *c* in *e* sit *m*.
Sed quia *h* est ut *b* in *e* et *c* in *e*. Sed *b* in *e* et *b* in *d* est
ut *b* in *de*, erit *gh* tamquam *m* et *de* in *b*. Et quia *de* in *b*
et in *c* est ut *l*, erit *ghk* ut *lm*. Sit igitur differentia *g*
10 et *m*, *n*. Qui quoniam dati sunt, erit *n* datus. Sed *hk* est
datus ad *a* et *l* datus ad *e*. Semper ut differentia *g*. Et igitur vel *hk* cum *n* dato facit *l*, vel *nl* erit ut *kh*; utrobique
autem datus est. Verbi gratia: Esto unus numerus radici duplus, alter triplus. A triplo autem demantur vi et a duplo
15 iiii, et residuum in residuum faciat cl. Sed duplum in triplum facit sescuplum quadrati et vi in iiii xxiiii. Iterim
autem triplum in iiii facit duodecuplum radicis et vi in duplum similiter duodecuplum. Itaque sescuplum quadrati et
xxiiii sunt ut radix vigiesquater et cl. Demptis ergo xxiiii
20 de cl remanebunt cxxvi et radix vigiesquater ut quadratum
sexies. Quare quadratum semel est ut radix quater et xxi.
Quadratum ergo dimidi iiii, quod est quatuor, addatur ad
xxi et erit xxv, cuius radix v, cui addatur dimidium quatuor
fient vii et ipse est radix.

IV-30.

SI VERO AB ALTERO DETRAHATUR NUMERUS DATUS, UT EX
RESIDUO IN RELIQUUM FIAT NUMERUS DATUS, ET SIC QUOQUE
IDEM DATUS ERIT. Sint ad *a* dati *bc* et *d*, et ex *b* in *d* fiat
e datus, posito quod *c* sit datus. Ex *bc* autem in *d* fiat *ef*
5 et sic *f* fiet ex *c* in *d*. Atque *ef* datus est ad *z* et *f* datus
ad *a*. Quia ergo numerus datus est ad quadratum, constat ex
numero dato et numero ad radicem dato, erit radix data. Verbi
gratia: Unus est ei duplus et alter ei duae quintae, et de

duplo sublatis iiii residuo in alterum ducto fient xii. Si
10 autem totus duplus in illum ducatur, fient iiii quintae quadrati et iiii in ipsum facit viii quintas radicis. Igitur
iiii quintae quadrati sunt ut viii quintae radicis et xii.
Secundum operationem ergo praemissam erit radix v, quadratum
xxv.

    IV-31.

SI FUERINT DUO NUMERI AD UNUM DATI ALTERIQUE DATUS
NUMERUS ADDATUR ET AB ALTERO DETRACTO NUMERO DATO, SI POST-
MODUM EX SE DATUM NUMERUM FECERINT, NUMERUM ILLUM DATUM
ESSE CONVENIET. Sint ad a, bc et d dati et c datus ad bc,
5 e datus addatur d fiatque ex b in de, f datus atque ex bc
in de fiat gh, ut ex bc in d fiat g. Et ex bc in e fiat
h cui equalis sit *flm*, fietque *lm* ex *de* in c. Estque l
datus ad a et m simpliciter datus. Differentia autem l ad
h sit t data ad a. Aut ergo t cum *fm* dato est ut g, da-
10 tus ad z; aut *fm* datus est ut t datus ad a, et g datus ad
z. Quodcumque fuerit, a semper datus erit. Verbi gratia:
Unus ei sesqualter alter medius maior dematur unus, alteri
addatur tres et sic ex eis facit xxv. Sed sesqualterum in
dimidium et tria facit iii quartas quadrati et quadruplum
15 radicis cum dimidio. Item unum dimidium et tria facit eadem. Quare xxviii et dimidium sunt ut iii quartae quadrati
et quadruplum radicis cum dimidio. Quare xxviii est quantum iii quartae quadrati et quadruplum radicis. Itaque, secundum quod super in opere processit, erit radix iiii.

    IV-32.

SI VERO EISDEM NUMERIS DATI NUMERI ADDANTUR VEL
DETRAHANTUR UT POST HOC PROVENIAT NUMERUS VEL AD QUADRATUM
VEL AD RADICEM DATUS, ET RADIX ET QUADRATUS DATUS ERIT. Si
quidem addantur dati numeri, ut si productum datum ad quadratum, detracto eo, quod fit ex uno in reliquum, remanebit
numerus ad quadratum datus constans ex numero dato et numero
ad radicem dato. Si vero ad radicem, detracto eo quod fit
ex quolibet eorum in numerum alii additum, remanebit numerus ad radicem datus constans ex numero dato et numero ad
10 quadratum dato. Datis autem numeris subtractis, si pro-

ductum sit ad quadratum, secundum demonstrationem xxix
huius detracto eodem producto ab eo quod fit ex uno in reliquum remanebit numerus ad radicem datus constans ex numero dato et numero ad quadratum dato. Si autem ad radi-
15 cem, fiet similiter numerus totus ad radicem datus constans
ex numero dato et numero ad quadratum dato. Verbi gratia:
In primo unus sesquitertius, alter duplus, minori quoque
duobus alii tribus additis ex ipsis, tunc proveniat numerus continens quadratum sexies, de quo quod fit ex uno
20 in alterum dempto, remanet triplum quadrati et tertia
tamquam octuplum radicis et vi. Secundum hoc ergo radix
erit iii et quadratus ix.

    IV-33.

QUOD SI ALTERI DATUS NUMERUS ADDATUR VEL DETRA-
HATUR DEINDE IN RELIQUUM DUCTUS PRODUXERIT NUMERUM QUAD-
RATO VEL RADICI DATUM, RADIX CUM QUADRATO DATA ERIT. Detracto sive addito dato numero, si productus ad quadratum
5 datus fuerit, quoniam numerus in alterum facit numerum ad
eundem datum, quod fit ex dato numero in reliquum, erit eidem datus. Sed radici datus erit. Quare radix quadrato
data. Igitur addito vero numero dato si productus fuerit
ad radicem datus, erit quod fit ex uno in alterum datum ad
10 radicem. Similiter etiam detracto. Quare radix ad quadratum data et ob hoc data. Verbi gratia: Unus sit sesquitertius radici, alter eius tertia et maiori addantur duo,
ut productum ex composito in reliquum sint duae tertiae
quadrati. Sed tertia in unum et tertiam facit iiii no-
15 nas quadrati. Ergo tertia in duo facit eius duas tertiae
radicis duae nonae quadrati. Radix ergo tertia quadrati
erit radix ternarius.

    IV-34.

SI VERO DATO NUMERO UNI ADDITO ET AB ALTERO DE-
DRACTO POSTMODUM FECERINT NUMERUM AD QUADRATUM VEL AD RA-
DICEM DATUM, QUADRATUS CUM RADICE DATUS ERIT. Si productus ad quadratum datus fuerit, atque alter cum sibi addito
5 in detractum a reliquo faciat numerum ad radicem datum, et
numerum datum erit, vel ipse cum sibi addito in reliquum

datum faciat numerum ad quadratum datum et numerum ad radicem datum, et tunc numerum datum. Sed reliquus totus in
illum et additum ductus facit totum numerum ad quadratum
10 datum et aliam ad radicem datam. Minoribus igitur de mai-
oribus detractis, vel relinquitur equalis dato ad radicem
vel ad quadratum datus, vel numerus datus equalis numero ad
radicem dato et numero ad quadratum dato, vel numerus ad radicem datus equalis numero dato et numero ad quadratum dato,
15 vel numerus ad quadratum datus equalis numero dato et numero ad radicem dato. Quodcumque fuerit, erit radix et
quadratus datus. Si productus ad radicem fuerit datus,
erit eadem ratione ducto toto in totum numerus coniunctus
ad radicem datus cum numero dato tamquam numerus ad radi-
20 cem datus et numerus ad quadratum datus. Demptis ergo
omnibus vel erit numerus datus equalis numero ad quadratum
dato cum numero ad radicem dato vel numerus ad quadratum
datus tamquam numerus datus et numerus ad radicem datus.
Et sic etiam radix data erit. Verbi gratia: Sit unus nu-
25 merus radici duplus, alter triplus et duplo addatur iiii et
reliquo detrahatur vi et productus postea contineat quadratum ter et eius tertiam. Quia autem duplum in vi facit duodecuplum radicis et iiii in vi xxiv. Quia item triplum in
duplum facit sextuplum quadrato et triplum in iiii duodecu-
30 plum, erit ut sextuplum quadrati et duodecuplum radicis sit
ut quadratum triplum et tertia et duodecuplum radicis et
xxiv. Dempto igitur utrobique duodecuplo radicis itemque
tribus et tertia ablatis de vi remanet duplum et duae tertiae
quadrati ut xxiv, quadratus vero ix. Radix autem tres et
35 sic in ceteris procedit operatio.

    IV-35.

SI NUMERUS AD QUADRATUM DATUS FECERIT NUMERUM AD RA-
DICEM DATUM, RADIX EIUSDEM DATA ERIT. Radix sit *a*, quadratus sit *b*, numerus ad eum datus *e*, qui in se ductus faciat *g*
datum ad *a*. Sit autem *d* quadratus *b*. Erit igitur inter *b*
5 et *d* numerus proportionaliter medius secundum proportionem
b ad *a* et ipse sit *c*. Et quia *a* in se facit *b*, et *a* in *b*
faciet *c*, et *a* in *c* faciet *d*. Et quia *g* ad *d* datus est,

erit et *a* ad *d* datus, quare *c* datus, qui est cubicus *a*, cuius
latere cubico extracto, proveniet *a* datus. Verbi gratia:
10 Dimidium quadrati in se facit numerum, qui radicem quin-
quagies quater continet. Sed dimidium in se facit quartam quadrati ipsius quadrati et quater liiii facit ccxvi.
Igitur ducentorum xvi latus cubicum extrahatur et provenient vi et ipse erit radix quadrati, qui est xxxvi.

EXPLICIT LIBER QUARTUS XXXV PROPORTIONES CONTINENS.
ET CUM EO FINIUNTUR DATA JORDANIS
SECUNDUM OPERATIONEM NUMERORUM.

Notes to the Critical Edition

Titulus. *om M, D, J*; DATIS *om P.*
I-1.
2. Etenim: Quia enim *M, B, J*; portio: proportio *M, B, J.*   4. *post*
differentia *add V* minoris.   5. sicut: sic *M, B, J.*
I-2.
DIVIDATUR *om P, R.*   4. differentiae *om B.*   16. vii, xi, xv: xx, xvi, xii
J. S.
I-3.
3. datus *om M, P, O, J*; in *om O, P.*   5. dempto *om B.*   6. *post* ab ad c
add W: Quod enim minor patet ex 3 secundi Euclidis. Subrahatur enim
c minor portio ad *ab*, relinquitur ergo differentia *ab* ad *c* sitque *am.*
Dividitur ergo numerus *ab* in duas partes in *am* et *b* et additur ei equalis
uni dividentium scilicet *c*, est enim *c* equalis *mb.* Age ergo ex 8 secundi
quadratum totius *abc*, hoc est numerus *e*, superficie quadruplum *d*, hoc
est numerus *f*, quadrato *am* minori, qui est differentia *ab* ad *c*, hoc est
numerus *g.* Ex hoc etiam haberi potest per 4 secundi, quod duo quadrata
ab et *c* pariter accepta valent duplum eius quod fit ex ductu *ab* in *c*, hoc
est medietatem *f* et quadratum differentiae *ab* ad *c*, hoc est numerus *g.*
Valet enim *e*, ut ostensum est, *g* et *f*, et valet idem *c* ex quinta secundi
duo quadrata pariter accepta et medietatem *f*, hoc est quadratum differentiae et duplam superficiem. Subtracta scilicet medietate *f* hinc inde,
quae communis est. (*Postea*: Extrahatur ergo radix *g* et sit *am.* Eritque
am differentia *ab* ad *c.* Cumque sit *abc* datum, erit et *c* et *ab* datum ex
prima.)   7. sit *b*, eritque *b*: *h* (bis) *J*; *b* datum: *bc* datum *B.*   8. *b*: *h J, A.*
13. dimidietur . . . iii: dimidetur iii *J, S.*
I-4.
2. DATUM MODO PRAEMISSO *B, D.*   3–4. *loco g* fuerit . . . de *g leget B g*
scilicet quadrata coniuncta. Subtrahendo quadrata parcium coniuncta
de quadrato totius numeri.   7. sublato *de om B.*   8. xlii . . . remanebunt *om B.*   2–6. *loco* Modo . . . data *leget W*: Sint duo quadrata
pariter accepta *ab*, sitque quadratum totalis propositi *ac*, maius enim
est, ut praeostendimus. Cum ergo *ac* datum sit, et *ab* datum, erit et *bc*
datum, quod per 4 secundi dupla est superficies. Cum ergo medietas eius
data, et omnia per praemissam erunt data.
I-5.
1. IN DUO *om P.*   3–8. *loco* Maneat . . . datus est *legel W*: Cum sit
superficies data, et quadruplum superficiei sitque *ab* et addatur *ab* et totus
ac datus est, qui, ut in 3 praesentis ostensum est, quadratum est totalis

prius non dati. Quare, cum est radius eius, habemus propositum.   4. *b*:
1 *M, B*; *b*: *h J, A*.   7. *post f add D* qui ex decima Arismetice Iordani
quadratus; *f*: *e J, A*.

I-6.

5. totius: divisi *M, P, D, J, A, B*.

I-7.

5. *c* ad *ab*: *c* ad *b O*.   12. cxcvi: clxxxvi *J*.

I-8.

4. *post* numeri *add D* ex decima octave Arismetice.   7. viii: ato *J*.

I-9.

4. *post* qui *add B* cum.   9. *post a add B* erit *df* et.

I-10.

3. *post* NECESSE *add B* ESSE.

I-11

2. IN RELIQUUM *om M, J, A, S*.   5, ipsum: duplum *D*.   6. Si: Sed *B*.

I-12.

1. CUM *om M*.   2. DIFFERENTIAE *om M*.

I-13.

1. ALTERIUS: UNIUS *J*.   4. data, sunt etiam et: data sunt, data sunt *J*.
5. ductum: duplum *J*.   6. quae duplata: duplata *J* ; ut : notum *P*, non
R.

I-14.

2. ET QUADRATO: DE QUADRATO *J*.   4. bis: *b B, om J*.   5. ipsum *om J*.
11. exit: erit *J*.   12. *post* maius *add B* dividentium vii et iii.

I-15.

4. minus: numerus *P*; *post* ipsius *add D* quibus demptis de numero
relinquitur quadratum differentiae cum duplo ipsius.   10. *post* duplum
add J huius.   14. dividentia vii et iii *om B*.

I-16.

18. et si hoc: et vero si *J*.   21. excedit v: excedit. Et excedit v *J*.

I-17.

1. DATO NUMERO IN DUO DIVISO: ALTERO NUMERO DIVISO *R*. EXIERIT:
EXIGERIT *B* (Ista litterarum ordinatio postea occurit; e.g. I-18, I-19, sed
non in I-20!).   6. *post* numeri *add D* non totius sed secundarii.   11. et
fiant . . . xlviii *om O*; ī̄.ccciiii: ī̄.ccxiiii *B*, ccxiiii *J, S*; ī̄.dcciiii: ī̄.dccxiiii
B, ī̄dcc *J, S*.

I-18.

5. *e*: *f B*; *f*: *e B*.   5–6. diviso . . . quadrata *om B*.   8. *e*: *d J*.   9. *post*
1 *add D* per tertiam huius.   13. a: de *J*.

I-19.

4. tunc: et *c D, B*.

I-20.

4. *f*: *p J*.   5. facit *om O*; *g*: *d J*.   6. sicut *a* ad *b*: sicut *g* ad *b O*; quare
e: quare *ef J*.   8. Quare *g* ad *ab*: quare *g* ad *b O*.

I-21α.
1. DATO NUMERO IN DUO DIVISO: ALTERO NUMERO DATO *R*.   3. Cum *c* numerus datus per *a* et: E numerus datus et *R*; Cum *c*: Cuius *V*.   6. fit: sicut *V*.   7. ex *a* in *b*: ex *ab B*; in *b om R*; *post h add D* provenit enim utrobique multiplex numerus, scilicet denominatus ab *ab*, si quis subtiliter inspiciat.   8. Ideo ducatur *ab* in *c om R*.   11. xl: et . . . *V*.   13. *post* reliquum *add D* ut ex II in VII.

I-21β.
3. *post* FUERIT *add J, S*   ET.   4. *a, b om J, S*.   6. *c*, et: *c*, in *J, S*.   7. *e*: *d J*.

I-22α.
1. *post* UNIUS *add D* EXEUNTIUM.   3. igitur *a*: igitur *h B*.

I-22β.
3. exeat *e*: exeat *d M*.   4. *ec*: *ac W*.

I-23.
4. quodlibet: quia *D*.

I-23*.
3. ERUNT . . .: IPSA ETIAM DATA ERUNT *W*.   6. erit et: atque *W*.   10. denominata . . . autem: ducta in reliquam. Cui *W*. (*J* exemplar est huius propositionis singularis.)

I-24.
6. semper: super *B*.   12. *ad*: *ade O*.   12–14. quod cum . . . dimidium ad *om B*.   13. *g*: *ge J, S*.   14. omnia erunt nota: differentia erit nota *D*; *ad*: *ade J, S*.   15. in (dimidium) *om O, P*.   17. contingit: erit *J, S*.   22. xlvi: 44 *B*.

I-25.
4. DATA *om B*; SINGULA . . . EST: SINGULA SUNT DATA *J, S*.   8. *c*: se *O*.   9. *e*: *ae J, S*.   10. ob hoc: *ab* et *b* et *J, S*.   14. ubi incidit: ut modo *J*.   16–18. dimidii . . . reliquo *om O*.

I-26.
8. *post* datum *add D* quia, quota pars *e* est *a*, tota pars *g* est *l*, sed hoc est secundum *e*—quod autem sit sic, ut dixi, patet intuenti; Ut solet *om J, S*.

I-27.
6. *post* cum *add J* sit *c*.

I-28α.
1. QUOD: QUODCUMQUE *D*.   3. et *k om B*.   10. solet x *om R*; x *om V*; ⟨vi⟩: iiii *P, O, R, V*.

I-28β.
(Haec est propositio 21 Alpha-familiae.)   4. *c om J*.   6. ⟨quantum⟩: quadratum *M*.

I-29.
8. se *om B*.

I-EXP.

Explicit . . . proportiones *om* P, O, B, J, S; *habet* V finit primus Jordanis de datis. Incipit secundus.

II-INC.

om P, O, B, J.

II-1.

2. altera: alterna *M*.

II-2.

7. idem: ipsum *J*.   14–16. xxvi . . . est *om J*.

II-3.

5. Per duo ergo et duas tertias *om O*.

II-5.

3. proportio *om J*.   6. *post* unum *add J* per duas.   7. unum *om J*.

II-6.

5. quorum: quare *M*.

II-8.

3. coniungantur *om J*.   6. bis: tres *J*.   7. *post* tertium *add J* et quartum.

II-9.

3. ipsi: ipse *J*.   4. ad compositum: ad eundem compositum *J*.

II-10.

4. *ab*, *cd*: abc *A, S*.   5 *a om M, P, C*.   8. *post* ad *g add D* subtracta enim minore proportione a maiore remanebit proportio *a* ad *g*, quae est differentia.

II-11.

2. ALIO DATO *om omnes codices excepto O*.   10. fiat: si ad *P*. 13. minorem: maiorem *O*.

II-12β.

4. ⟨*d*⟩ ad se: c at se *M*. *d*: *c J*.   6. *cg*: *g J*.

II-13α.

5. *a* et *c* sit *g*: *b* et *d* sit *g O*.   6. minor: maior *P, O*.

II-13β.

11. *post* tamquam *add J* illud.

II-14β.

3. duo *om J*.   6. *c*: *d M, J*.   13. iii *om J*.

II-15.

1. *post* ET *add D* TOTUM.

II-16.

5. omnibus: duobus *P*.

II-17.

8. tertia: quarta *M, O, P, R*.

II-18.

Omnes codices positiones demonstrationis et exempli inter se mutant. Insuper, nomen "demonstratio" probationem introducit eorum.

II-20.

6. igitur ad: igitur *e P, M.*   12. iii: iiii *P, O, M.*   18. bis: ter *D.*
23. *post* cetera *add D* eius duas septimas et duas vigesimas primas, et
primus.

II-21.

1–2. VEL NUMERUS AD IPSUM DATUS *om M.*   5. est differentia. Et quia
P, O; nota: aliquid in P deletum videtur *c ad h* ut in *M* esse.

II-22.

5. datus *om O.*   6. data *om O.*

II-23.

2. SEQUENS: PRAECEDENS *D.*   13. et *b*: ad *O.*   15. quare in *de om D.*
16–17. et sic in *ba.* Erit itaque *om D.*

II-25.

26–27. cxix et medietas secundi *om O.*   34. cxx: dcc *O.*   36. cxix:
xcxix *O.*

II-26.

7. *mno: mo O.*   10. *ah: ab O.*   12. *v: n B.*   13. *mo: mc O.*   14. quia
pg: quia *b O.*   20. *nd: no d M.*   21. ad *a* et datum *om O.*   33. tribus
de tribus et quarta: triplo tertii et triplo quarti de triplis et quartis eorum
D.   52. xxiv: xv *O.*   54. xxiv: ix *O,* xx *M.*   55. vigesimas octavas:
vigesimam primam *M, O.*

II-27.

11–12. quod . . . sic: quod ipsi nominarunt ad xii in positione esse falsi
D.   16. trium: xii *D.*   23. cum *b* et: erit *e D.*   35. *e: d D.*

II-28.

1. Solummodo *P* hanc propositionem eiusdem litteris mensurae ut supra
incipit; ideo, ista numerum habet metipsum.   13. xxviii: xxiiii *M, O, P,
D.*   27. *post* inventa *add B*: In ista autem operatione vero erit 84 quasi
numerus datus et quod aggregatur ex ipso et primo invento, scilicet $84\frac{1}{9}$,
sicut patet per operationem et per rationem. Et istius aggregati ad primum
propositum inventum considerabitur proportio, scilicet $87\frac{1}{9}$ ad $3\frac{1}{9}$, et est
vigecupla octupla, et talis erit 28 ad primum suorum, scilicet ad unitatem.
Et hoc manifeste docetur in opere partium, quo utuntur Arabes. Com-
parat $\frac{11}{6}$ et . ? . tamquam numerum datum inventum quia compositum
ex $6\frac{1}{6}$ ad sextum quod est loco primi, ut sic inveniat proportionem 37 ad 1.
(*Nota:* mutatio a numeris romanis ad arabios subjicit additionem glossam
esse.)

II-EXP.

xxviii: xxvi *M.*

III-4.

2. PROPORTIO *om P, M.*

III-5.

4. quantum ex *a*: quantum *a O.*

III-7.

5–6. et *e* ex ductu *a* in *c om P, M, O*; *e om P, M, O*.   10. quater: quartum *O*.

III-9.

1. MEDII: MEDIUS *P*; *post* UNO *add P, O* DATO.

III-10.

4. *post* enim *add D* extremorum.

III-11.

15–17. duarum . . . tertia *om O*.

III-12.

13–14. et ex medio . . . sit xxiiii *om O*.

III-13.

4. dupliciter: duplicatum *O*.   25–26. esse medium: esse dimidium *O*, *P*.

III-17.

7. *a* ad ⟨*c*⟩: *a* ad *b O, M*.

III-18.

7. quaecumque: cumque *D*.

III-19.

12. xxx, atque: xxx, ad *M*.

III-20.

20. tertius vi *om M*.

III-21.

4. dato sit *e*: dato. Si est *M*.   6. *post* differentia *a add M* ad *d* constat ex differentia.

III-22.

12. Itaque medium *om O*.   13. continet: continens *P*.

III-23.

7. ad proportionem primi ad secundum *om O*.   15. et diminuendo et ita ad extremos: et continuando ad extremos *M*.   22–23. tertiam: quartam *O*.

IV-1.

4–5. datum ad *d*: datum ad *V*.

IV-2.

4. sit: est *J*.   6. ducantur, et *om V*; alii: alius *V*.

IV-3.

2. ATQUE DIVIDENTIUM DIFFERENTIA *om P*.   5. Sitque: Sintque *M*.
6. quare *f* ad *g om V*; sicut *h*: sicut *J*.   10. *b*: *d J*.   12. numeri *om J*.

IV-4.

3. est: *e J*.   4. *l* per *d*: *l* per *b P*.   5–6. differentia *b* et *c*: differentia *e* et *b O, V*, differentia *a* et *b M*.   6. ex *c* in *b*, erit *c* et *b*: ex *a* in *b* erit *a* et *b M, J*; erit *c* et *b*: erit *a* et *b O, V*; erit *a* et ? *P*.   9. xx *om O*.   13. xiii: xvi *J*.

IV-5.
6. sesquitertius: sexquitertius *V*.
IV-6.
4. *c* ad *d*: *c d J*.
IV-7.
4. addit: accedit *V*.   6. hoc *om P, M*; ob hoc latus: et abiatur *J*.   8. sit: et *J*.   10. Quare *om M*.
IV-8.
4. *c*: *cd J*.   6. *d*: *g M, J*.   9. Sit: Esto *V*, est *J*.   13. ablatis: sublatis *M*.
IV-9.
4. *c*: *e P, O, D*; *dc*: *de O, M, B, D, V*.   9. remanere . . . potest *om V*.
12. per vi: per se *V*.   13. iii: 6 *V*.
IV-10.
2. DATO: QUADRATO *M, J*.   4. *e*: *c V*; *post* sit *add J* ad.   7. *b* in *g* et *om J, V*; *d*: *g J*; *post d* in se *add V, J* et *gcd* in *g*.   11. quadrato: 4ti *V*.
12. *post* dimidii *add J* per.
IV-11.
4. *b*: *bc J*.   5. ita sit *g* ad *c*: ita *a* fit *g* in *c O*.   6. *bd*: *bg J*.   7. in *e*: in *c O, V*.
IV-12.
5. *d*: *c J*; *bc*: *be M, V*.   11. ea: eadem *J*.   13. īīd: 250 *V*; cxxv: cxx *P, D, M, B*, xx *O*; quae sunt *om J*.
IV-13.
4. *cd*: *cb J*.   5. datus *om V*.
IV-14.
2. EX ILLIS *om M*.   4. datus *om J*.   12. v *om O*.
IV-15.
2. SIMILITER: SIMUL *M, D, R*; ILLIUS DATA *om M*; *post* SIMILITER *habet J* AD QUADRATUM ILLA QUOQUE SIMUL AD IPSUM DATA ERUNT.   4 et 5. *t*: *c P, J* (4 solus).   5. *e* ad *d*: *e V*.   6. et *f* et *g*: et *fg V*.   8. additus: addens *O*.   9. v nonas (et reliqua tractatus!) *om Alpha familia MSS*.
IV-16.
2. UNO: ALTERO *J*.   10. est *om J*.
IV-17.
1. SIMUL: SIMILITER *J*.   7. illi: illo *V*.   9–10. cuius . . . duo *om J*.
IV-18.
10. fient: faciunt *J*.
IV-19.
3. quodque: et quod *V*.   4. *e*: *d J*; *d*: *a J*.   7. ⟨v̄⟩cccclxxvi: īīīīcccclxxvi *M*, 4486 *V*; cuius . . . xxvi *om J*.   8. eo: eis *V*; dlxxvi: lxxvii *M, J*.
IV-20.
3. *c* ad *d*: *c* ad *g M*.   3–5. proportionaliter . . . secundus: proportionaliter est ut *c* ad *d* et *e* ad *c* sicut *b* ad *e* quare *e* sit primus et quartus dati et

c et *d* secundus *V*.   4. *post* proportionaliter *add J* est; erit *c* . . . *d*: erit *c* ad *d* sicut *d* ad *e J*; Quare *c*: Quare *e* sit prima, quae fiat dati. Et *e* et *d* secundi et tertius *J*.   7. sit: est *V*.   8. *e*: *d J*.   10. ccxxv: cxxv *J*.   12. xxxiiii: xxxvii M, xxxvi *J*; *post* mediato *add J* fient.

IV-21.

5. *post* ipsa *add J* data.   6. cccxxiiii: ccxxxiiii *J*.   9. quater *om J*.

IV-22.

1. SIMULQUE: SIMUL *W*.   6. est *om M*; c: *e M*; ⟨duorum⟩: duo *M*, *V*, *J*.

IV-23.

1–9. SI ET . . . vero x *om F*.   2. NECESSE EST: CONVENIT *V*.   7. xvii: xviii *M*; componitur: componetur *M*, *V*.

IV-24.

2–3. numeri *om V*.   5. *e* ad *b*: *b* ad *e J*.   9. *e om J*.   10. *e*: *g M*. 11. *e*: *g J*.   15. xviii: 30 *V*; *post* quadratum *add M*, *V*, *J* et dimidium.

IV-25.

7–9. atque . . . *l* ad *e om J*.   9. igitur *l*: igitur *b M*.   11. *l* in se: *a* in se *V*.   12. Sed in *e om V*.   16. differentiae: totius *V*.   19. eidem: ei *M*. 23. xxviii: xxv *M*, *J* (*idem* 25 *V*).   26–27. in quae dividitur x et duae tertiae, quia viii *om V*.   29. sumantur *om M*; 1: 5 *V*.

IV-26.

7. *post* Itaque *add V* ab.   8. et *e*: et *c V*.   10. et *a* et *b*: et *ab V*; *post* sit datus *add J* et *ab* datus.

IV-27.

4. addiciantur: addantur *V*.   5. *b* in *c*: *b* in *d M*, *F*, *V* (*at* non *W*!). 10. quadratum *om V*.   12. xx: 21 *V*.   13. in: ad *V*.   14. radicis: quadrati *M*, *V*, *J*.   16. duodecima quadrati. Ergo: duodecima. Quadrati ergo *M*.   17. septem: 8 *V*.

IV-28.

1. IN *om V*.   3. in *c om J*.

IV-29.

3. PROVENIET *om J*.   5. bc: de *V*.   6. et *om M*.   10. *m*, *n*: *m* et *n V*, *J*. 11. datus ad: datus et *J*; ad *a*: et *a M*; ad *e*: ad *z V*; semper . . . g *om V*, *J*.   13. *post* est *add J a*.

IV-30.

5. *f om V*.   6. est *om M*; ex *om M*.   8. alter ei: alter eius *V*.   12. iiii *om V*.

IV-31.

3. EX . . . FECERINT: EX SE SECUNDUM NUMERUM DATUM *J*.   4. et *d om V*. 4–5. ad bc, *e*: a bc et *e V*, *J*.   5. de, *f*: d, ef *J*.   6. de: ge *M*, *J*.   9. a: h *J*.

IV-32.

7. ad radicem *om J*; *post* dato *add J* et numero ad radicem dato.   8. eorum *om M*.   11. xxix: xxxix *M*.   12. fit *om J*.   14–16. Si autem . . . quadratum dato *om V*.   19. sexies *om V*.

122

The Critical Edition

IV-33.

2. DUCTUS: PRODUCTUS *J*.   15. eius duas tertias: eius duas nonas resid-
uas. Sed tertia faciat duas tertias radicis. Erunt duae tertiae *J*.

IV-34.

4. atque: cumque *J*; sibi: sit *V*.   5. datum: totum *V*.   6. vel: ut *J*.
9. totum *om V*.   25. numerus *om J*.   30–31. sit ut . . . radicis *om J*.

IV-35.

3. *b*: *d V*.   11. quater *om J*.

The English Translation

Introduction to the Translation

Let the translation give the thought
of the author in the idiom of the reader.
ANON.

Two goals were set for the translation and a third was born of necessity. The two were to make it mathematically correct and to render it readable; the third was to give the reader the experience of rhetorical algebra. Rugged as the experience will be, some of the roughness has been eased by removing the rocks of literalness. For example, in I-1 are the sentences "Sublata ergo differentia de toto, remanebit duplum minoris datum. Qua divisa, erit minor portio data. Sicut et maior." Terse as the Latin is, the meaning is clear. The English may be rendered in two ways, (1) literally and (2) liberally.

(1)	(2)
The difference therefore subtracted from the whole, there will remain given the double of the smaller. Which divided, there will be given the smaller portion. And so the larger.	Subtracting therefore the difference from the total, what remains is twice the smaller. Halving this yields the smaller and, consequently, the larger part.

My taste tends to the liberal. Those who prefer a more literal translation may well exercise their skill.

A comment is necessary about my rendering of literal (alphabetic) numbers used by Jordanus. He always begins with the initial letters of the alphabet and works his way, as far as need demands, through the alphabet. For the sake of clarity to the modern reader, I nearly always employ the Cartesian convention. That is, the letters at the beginning of the alphabet are known numbers or constants; those at the end are unknown numbers or variables. The few exceptions to my procedure are found in the β-set of propositions in Book I and in a few other propositions where Jordanus nearly exhausts, or actually uses, all the letters of the alphabet.

The footnotes attached to the statements of propositions constitute *apparatus fontium*. It is meant to be suggestive rather than definitive. Jordanus quoted no sources. Lacking first-hand evidence, any attempt at identifying sources betrays the bias of the investigator. My bias is that Jordanus was well read in mathematics and natural philosophy; witness the breadth of his writings. I believe that he drew on his broad knowledge and creative ingenuity to fashion propositions he thought practical. I suspect that he was so steeped in the literature available, much of which was discussed above under the rubric "Sources," that the ideas—his understanding of them, both new and old—flowed from his pen effortlessly. Regardless, this set of footnotes refers to possible sources of the propositions. The fonts are: Euclid's *Elements* and *Data*, al-Khwārizmī's *Liber algebre*, abū-Kāmil's *Kitāb fī al-jābr wa'l-muqābala*, and Fibonacci's *Liber abaci*. All of these, at least, were available to Jordanus.

Phrases such as "by I-1" are used extensively in the footnotes. They mean that the problem-solver must refer to Book I, proposition 1, to complete the solution. The reader will eventually note that some of the phrases are not the same as those I have in the fourth column for my symbolic translation. The reason for the difference is that my third column, *Reduction to Canonical Form*, sometimes makes the transition to ultimate step in order to show the canonical form.

Book One

Definitions.

1. *A number is given* whose quantity is known;
2. *A number is given in relation to another* where the ratio of the one to the other is given;
3. *A ratio is given* whose denomination[1] is known.

I-1. IF A GIVEN NUMBER IS SEPARATED INTO TWO PARTS WHOSE DIFFERENCE IS KNOWN. THEN EACH OF THE PARTS CAN BE FOUND.

Since the lesser part and the difference equal the larger, the lesser with another equal to itself together with the difference make the given number. Subtracting therefore the difference from the total, what remains is twice the lesser. Halving this yields the smaller and, consequently, the greater part.

For example, separate 10 into two parts whose difference is 2. If that is subtracted from 10, 8 remains, whose half is 4. This is the smaller number and the other is 6.

I-2. IF A GIVEN NUMBER IS SEPARATED INTO AS MANY PARTS AS DESIRED WHOSE SUCCESSIVE DIFFERENCES ARE KNOWN, THEN EACH OF THE PARTS CAN BE FOUND.

Given is the number a which is divided into w, x, y, and z the least of the parts. Since the successive differences of all these are given, each difference can be expressed in terms of the difference of each number with z. Therefore let f be the difference of w and z, and the sum of g and h be the sum of the differences of x and z with y and z. Now because z makes each of those equal to each of these, it is obvious that thrice z with the sum of f, g, and h equals those three. Therefore four times z with the sum of f, g, and h equals a. Subtracting this sum from a leaves four times z, which is now known; hence z is found. By addition of differences the other parts are found.

For example, separate 40 into four parts whose, successive differences are 4, 3, and 2. Therefore the difference of the first and last is 9, of the second and last is 5, and of the third and last is 2. Their sum is 16. This subtracted from 40 leaves 24 whose fourth is 6, the least of the four parts. By adding this to 9, 5, and 2, the other three parts are found, namely 8, 11, and 15.

I-3. IF A GIVEN NUMBER IS SEPARATED INTO TWO PARTS SUCH THAT THE PRODUCT OF THE PARTS IS KNOWN, THEN EACH OF THE PARTS CAN BE FOUND.[2]

Let the given number a be separated into x and y so that the product of x and y is given as b. Moreover, let the square of $x + y$ be e, and the quadruple of b be f. Subtract this from e to get g, which will then be the square of the difference of x and y. Take the square root of g and call it h. h is also the difference of x and y. Since h is known, then x and y can be found.

The mechanics of this is easily done thus. For example, separate 10 into two numbers whose product is 21. The quadruple of this is 84, which subtracted from the square of 10, namely from 100, yields 16. 4 is the root of this and also the difference of the two parts. Subtracting this from 10 to get 6, which halved yields 3, the lesser part; and the greater is 7.

I-4. IF A GIVEN NUMBER IS SEPARATED INTO TWO PARTS THE SUM OF WHOSE SQUARES IS KNOWN, THEN EACH OF THE PARTS CAN BE FOUND.[3]

As in the previous method, call the known [sum of the squares] b, and [let] e—twice the product of the two parts—[be found by subtracting the sum of the squares from the square of the given number—*trans*]. Subtracting e from b yields h, the square of the differences,[4] whose root is c. Hence, all the parts can be found.

For example, separate 10 into two parts, the sum of whose squares is 58. Subtract this from 100 to get 42, which in turn is subtracted from 58 to yield 16. The root of this is 4, which is the difference of the parts. As before[5] these are found to be 7 and 3.

I-5.   IF A NUMBER IS SEPARATED INTO TWO PARTS WHOSE DIFFERENCE IS KNOWN AND WHOSE PRODUCT IS ALSO KNOWN, THEN THE NUMBER CAN BE FOUND.

As before [6] let a be the known difference of the parts and b be their given product. Let h be the square of the difference of the parts and e be four times their product. Call the sum of these f. f is equal to the square of $x + y$.[7] And so $x + y$ is found.

For example, let the difference of the parts be 6 and their product, 16. Twice this is 32, and twice again is 64. To this is added 36, the square of 6, to make 100. The root of this is 10, the number that was separated into 8 and 2.[8]

I-6.   IF THE DIFFERENCE OF TWO PARTS OF A NUMBER IS KNOWN AND ALSO THE SUM OF THEIR SQUARES, THEN THE NUMBER CAN BE FOUND.

Subtract from b, the given sum of the squares, h, the square of their given difference. Call the remainder e, which is also twice the product of the parts. The sum of e and b is the square of the desired number, f. Hence, take its root to find the number.

For example, let 68 be the sum of the squares from which 36, the square of their difference, is subtracted. The remainder, 32, is twice the product of the parts. Adding 68 and 32 to get 100, take the root of this to get 10. This is the desired number whose parts are 8 and 2.[9]

I-7.   IF ONLY ONE OF TWO PARTS OF A NUMBER IS KNOWN, PROVIDED THE SUM OF THE PRODUCT OF THE PARTS AND THE SQUARE OF THE UNKNOWN PART IS GIVEN, THEN THE NUMBER CAN BE FOUND.

Let the parts of the number be x and b, with b given. Also given is a, the sum of the product of the parts, and the square of x. Add z, equal to x, to $x + b$ so that the entire $x + b + z$ can be separated into $x + b$ and z. Now since $x + b$ times z equals the given a, and the difference of $x + b$ and z is the given b, then $x + b$ and z are found as are x and $x + b$.[10]

For example, let 6 be one of the parts and 40 the sum of the product and the square. Double 40 and redouble to get 160. Add to this 36 to obtain 196 whose root is 14. From this subtract 6 and halve the remainder to yield 4. This is the unknown part that with 6 makes the desired number 10.

I-8. IF A GIVEN NUMBER IS SEPARATED INTO TWO PARTS AND THE SUM OF THE SQUARE OF THE LESSER PART AND OF THE PRODUCT OF THE GIVEN NUMBER AND THE DIFFERENCE OF THE TWO PARTS IS KNOWN, THEN EACH OF THE PARTS CAN BE FOUND.

The square of the greater part is equal to the given sum.[11] Once its square root is found, then the other part can be found.

For example, separate 10 into two parts, and let the given sum equal 64. Its root is 8, which is the greater part, and the lesser is 2.

I-9. IF A GIVEN NUMBER IS MULTIPLIED BY THE DIFFERENCE OF ITS PARTS AND IS ADDED TO THE SQUARE OF THE GREATER PART TO MAKE ANOTHER GIVEN NUMBER, THEN EACH OF THE PARTS CAN BE FOUND.

Separate the given number a into x and y so that their difference is g. Multiply $x + y$ by g to get d. Square the greater part x to yield e. Thus the sum $d + e$ is given. Now let the square of $x + y$ be f. Thus the whole $d + e + f$ is known. Since $x + y + g$ is twice x, $d + f$ will be twice the product of $x + y$ and x. Likewise $d + e + f$ will be the sum of x squared and of twice x times $x + y$. Since $d + e + f$ and twice $x + y$ are known, so is x found together with y.[12]

For example, multiply 10 by the difference of its parts and add it to the square of the greater to make 56. Add this to 100 to make 156, which doubled then redoubled is 624. Add to this the square of 20, which is twice 10, to reach 1024. The root of this is 32, from which 20 is subtracted to yield 12. Half of this is 6, the greater part of 10, and the lesser number is 4.

I-10. IF THE SUM OF THE SQUARES OF THE PARTS OF A GIVEN NUMBER IS ADDED TO THE PRODUCT OF THAT NUMBER AND THE DIFFERENCE OF ITS PARTS TO FORM ANOTHER GIVEN NUMBER, THEN EACH OF THE PARTS CAN BE FOUND.

The given sum can be reduced to twice the square of the larger part. Hence, halve it and take its root to find the larger part.

For example, squaring and adding the two parts of 10 to the product of 10 and the difference of its parts yields 98 whose half is 49. The root of this is 7 for the larger part and the smaller is 3.

I-11.   IF THE SUM IS KNOWN OF THE PRODUCT OF A GIVEN NUMBER AND THE DIFFERENCE OF ITS PARTS AND OF THE PRODUCT OF ITS PARTS, THEN EACH OF THE PARTS CAN BE FOUND.
Now the given number is also equal to the difference of its parts added to twice the smaller part. Hence its square is the sum of itself times the difference of the parts and twice the product of the lesser and of the two parts. Now the product of the lesser and the whole equals the sum of the lesser times the greater and of the square of the lesser. If, therefore, from the square of the whole is taken the sum of the product of the whole and the difference and of the product of the two parts, what remains is the sum of the square of the lesser and of the lesser times the given whole. Hence, from what has gone before [13] the lesser and the greater can be found.

For example, the sum of the product of 10 and the difference of its two parts and of the product of the two parts is 89. Subtracting this from 100 leaves 11. Doubling and redoubling this yields 44, which with 100 makes 144 whose root is 12. The difference of this and 10 is 2 whose half is 1. This is the smaller number; the greater is 9.

I-12.   IF THE SUM OF THE SQUARES OF THE TWO PARTS OF A GIVEN NUMBER TOGETHER WITH THE SQUARE OF THEIR DIFFERENCE IS KNOWN, THEN BOTH PARTS CAN BE FOUND.
If the sum is subtracted from the square of the given number, what remains is twice the product of the two parts less the square of their difference. This becomes the sum of the squares of the parts less twice the square of their difference, and finally it is the given sum less thrice the square of the difference.[14] When, therefore, the remainder is subtracted from the given sum, take one third of what is left. The root of this is the difference that was sought. Hence, all can be found.

For example, square the two parts of 10, and adding them to the square of their difference yields 56. Subtract this from 100 to get 44, which in turn is subtracted from 56. The remainder is 12, whose third is 4. The root of this is 2, the difference of the parts. Therefore the larger number is 6 and the smaller is 4.[15]

I-13.   IF THE SUM OF THE PRODUCT OF TWO PARTS OF A GIVEN NUMBER
AND OF THE SQUARE OF THEIR DIFFERENCE IS KNOWN, THEN EACH PART
CAN BE FOUND.
By doubling the given sum you obtain the sum of the two parts squared together with the square of their difference. Since these are given, what is sought can be found.

For example, let the product of the parts and the square of their difference be 28. Doubling this produces 56, which are the three squares as above.[16] The rest follows as before.

I-14.   IF THE DIFERENCE OF THE SQUARES OF THE TWO PARTS OF A GIVEN
NUMBER IS KNOWN, THE PARTS CAN BE FOUND.[17]
By subtracting the known difference from the square of the given number what remains is twice the square of the lesser part together with twice the product of the parts. Halving this yields just the square of the lesser added to the product of the two parts; this is equal to the product of the given number and its lesser part. By dividing the difference by the given number, the lesser part emerges and the greater part can be found.

For example, take the difference of the squares of the two parts of 10, and let this equal 80. Subtracting this from 109 leaves 20, whose half is 10. Dividing 10 by 10 produces 1 for the smaller number, and 9 is the greater.

I-15.   IF THE SUM OF THE SQUARES OF TWO PARTS OF A GIVEN NUMBER
ADDED TO THEIR DIFFERENCE IS KNOWN, THEN THE TWO PARTS CAN BE
FOUND.[18]
If the sum is subtracted from the square of the given number and then the remainder is subtracted from the given sum, it is obvious that what remains is twice the sum less the square of the given number, that is to say, the product of the difference and itself increased by 2.[19] Since this is known, the difference can be found.[20]

For example, let the sum of the squares of the parts increased by their difference equal 62. Subtract this from 100 to obtain 38, which in turn is subtracted from 62. The 24 that results equals the square of the difference of the parts increased by twice the difference. Doubling and redoubling this produces 96. To this add the square of 2 to reach 100. Its root is 10. Take 2 from this and halve the remainder to obtain 4, the difference of the parts which are found to be 7 and 3.

I-16.   IF THE SUM OF THE PRODUCT OF THE TWO PARTS OF A GIVEN NUMBER AND OF THEIR DIFFERENCE IS KNOWN, THEN EACH OF THEM CAN BE FOUND. Separate the given number *a* into two parts, *x* and *y*. Set the sum of their product and difference equal to *b*. Doubling the sum gives *d*, which is to be subtracted from *e*, the square of the given number, to yield *f*.

If *f* is less than *d*, the difference must be found. If it is 4, then the difference of the parts equals 2; if it is 3, then the difference of the parts is either 3 or 1, and hence they cannot be unique.[21]

On the other hand, if *f* and *d* are equal, the difference of the parts is 4.

But if *f* exceeds *d*, set their difference equal to *g*. *g*, then, will be the product of the difference of the parts and the same difference less 4. Since this is known, the difference of *x* and *y* can be found.

For example, take the two parts of 9 and add their product to their difference to make 21. Subtract its double, 42, from 81 to get 39, which is 3 less than 42. Therefore[22] the difference can be 1 or 3, since both occur. It can be 1 if 9 is separated into 5 and 4; then 5 times 4 plus 1 equals 21. It can be 3 if 9 is divided into 6 and 3; then 3 times 6 plus 3 equals 21. So an error has happened! Try again.

Separating 9 another way produces the given sum 19, whose double is 38. If this is subtracted from 81, the remainder is 43. The difference between this and 43 is 5. Quadrupling this yields 20. Increase this by the square of 4 the double of 2 to get 36, whose root is 6. From this take 4, halve the remainder, and add it to 4 to make 5. And that is the difference of the parts which are 7 and 2.

I-17.   IF THE QUOTIENT OF THE PRODUCT AND DIFFERENCE OF THE TWO PARTS OF A GIVEN NUMBER ARE KNOWN, THEN THE PARTS CAN BE FOUND.[23] Since four times the product of the parts is equal to the difference of the squares of the sum and of the difference of the parts, it follows that the product of the difference and itself increased by 4 times the product equals the square of the given number. Hence the difference of the parts can be found.[24]

For example, set the product of the parts of 10 divided by their difference equal to 12. Quadruple this to obtain 48. Doubling and re-doubling 100 and adding it to the square of 48, namely 2304, the number 2704 is obtained. The root of this is 52, from which 48 is subtracted. Half of this is 2; and this is the difference of the parts.

I-18.   IF THE QUOTIENT OF THE SUM OF THE SQUARES OF THE TWO PARTS OF A GIVEN NUMBER AND OF THEIR DIFFERENCE IS KNOWN, THEN THE PARTS CAN BE FOUND.

Let the given number a be separated into x and y so that the sum of their squares is c and their difference is d whose square is e; and let the square of the given number be f. The quotient of c and d is g, whose double is $h + j$, also known. Because the squares e and f are twice c, it follows that the product of d and $h + j$ equals $e + f$. Note, however, that j equals d and j squared equals e; then j times h equals f, which is known. j and h are then found.[25] Consequently all the other values may be found.

For example, let the sum of the squares of the two parts of 10 divided by their difference equal 26. Doubling this yields 52, whose square is 2704. Subtract 400 from this to get 2304, whose root is 48. From 52 subtract 48 and halve the remainder to get 2, the difference of the parts.

I-19.   IF THE QUOTIENT OF THE TWO PARTS OF A GIVEN NUMBER IS KNOWN, THEN THE PARTS CAN BE FOUND.[26]

Let the given quotient of x and y be b. Increase this by one to get d. Because the product of y and b equals x, the product of y and d is equal to $x + y$. Dividing $x + y$ by d leaves y.

For example, let the two parts of 10 equal 4. Increase this by one and divide 10 by 5 to get 2, one of the parts.

I-20.   IF THE SUM OF THE QUOTIENTS OF THE TWO PARTS OF A GIVEN NUMBER EACH DIVIDED BY THE OTHER IS KNOWN, THEN THE PARTS CAN BE FOUND.[27]

Let the quotients of x and y be b and of y and x be c, and let one be added to b and to c to yield e and f. Let the product of x and y be g. Now since the product of x and f is $x + y$, and y times e is also $x + y$, it follows that e is to f as x is to y. Hence $e + f$ is to f as $x + y$ is to y, and alternately $x + y$ is to $e + f$ as y is to f. Since x times y plus f is equal to g plus $x + y$, then g is to $x + y$ as y is to f. Hence, g is to $x + y$ as $x + y$ is to $e + f$. Consequently the square of $x + y$ divided by the known $e + f$ produces g as known. Therefore x and y can be found.[28]

For example, briefly: let the sum of the respective quotients of the two parts of 10 equal $2\frac{1}{6}$. Add 2 to this to make $4\frac{1}{6}$, by which 100 is divided. The quotient is 26, which is also the product of the two parts.[29] Subtracting this four times from 100, as is customary,[30] yields 4, whose root is 2. This is the difference of the parts, which are 6 and 4.

I-21α. IF THE TWO PARTS OF A GIVEN NUMBER ARE DIVIDED INTO ANOTHER GIVEN NUMBER AND THE SUM OF THE QUOTIENTS IS KNOWN, THEN THE PARTS CAN BE FOUND.[31]

Divide the given number c by x and y and let the sum be given as $d + e$. Let f equal the quotient of c and $x + y$. $x + y$ squared shall be g. Then f times g equals the product of $d + e$ and h the product of x and y. Now the product of g and f equals the product of $x + y$ and c. Therefore by multiplying $x + y$ and c and dividing this product by $d + e$, h is found. Hence, x and y can be found.

For example, divide 40 by the two parts of 10 and set the sum of the quotients equal to 25. Then multiply 10 by 40 and divide the product of the two parts, namely, of 2 and 8.[32]

I-21β. IF A GIVEN NUMBER IS SEPARATED INTO TWO PARTS AND ONE IS MULTIPLIED BY A GIVEN NUMBER AND THE OTHER BY THAT PRODUCT AND IF THE PRODUCT IS SET EQUAL TO A GIVEN NUMBER, THEN THE TWO PARTS CAN BE FOUND.[33]

Let the parts be a and b. Multiply a by the given number c to get d, and multiply b by d to obtain the known e. Since this multiplication realized only as much as if b were multiplied by a and the product by c, then divide e by c to obtain what was given, namely the product of a and b. Since, therefore, a and b are found, so are $a + b$ known.

For example, multiply one part of 10 by 5 and the other part by this product to obtain 105. Divide all of this by 5 to obtain 21, which is equal to the one part of 10 multiplied by the other. Now, according to the custom, subtract 21 four times from 100 and the remainder will be 16. The root of this is 4 which is the difference of the parts. And these are 7 and 3.

I-22α.   IF THE PRODUCT OF THE QUOTIENTS FORMED FROM A KNOWN NUMBER
AND THE TWO PARTS OF A GIVEN NUMBER IS KNOWN, THEN THE PARTS CAN
BE FOUND.[34]

Let b be the product of d and e, and h be the product of x and y. Since the
product of x and y is h and the product of x and d is c, therefore c is to h
as d is to y. Likewise, since the product of e and y is c and the product of
e and d is b, then b is to c as d is to y as c is to h. Hence, by dividing the
square of c by b, one obtains h.[35]

For example, divide 40 severally by the parts of 10 and let their product be 100. Then divide the square of 40 by 100 to obtain 16. Proceed as
before.[36]

I-22β.   IF THE PRODUCT IS DIVIDED BY ONE OF THE PARTS AND WHAT
RESULTS IS GIVEN, THEN EACH OF THE PARTS CAN BE FOUND.[37]

Thus, divide d by b to yield the known e. Now since a is to e as b is to c,
therefore $a + b$ is to $e + c$ as a is to e. And so multiply e by $a + b$ and
divide this by $e + c$ to obtain a. Hence, b can be found.

For example, multiply one part of 10 by 5; divide this product by the
other part. The result is $7\frac{1}{2}$. Now multiply 10 by 5 to get 50.[38] Divide this
by $12\frac{1}{2}$ to yield 4, which is one of the parts.

I-23.   IF THE QUOTIENT OF THE PREVIOUS QUOTIENTS IS KNOWN, THEN THE
PARTS CAN BE FOUND.

As before [39] let c be divided by x and y to obtain the quotients d and e,
and divide d by e to obtain the known b. Now whatever is the product
of x and d equals y times e, each of which is equal to c. It follows that
x is to y as e is to d. Hence, the quotient of d and e equals that of y and
x, which is known. And so the parts can be found.

For example, let the two parts of 10 divide 40 so that the quotient
of their quotients is $\frac{1}{4}$. Hence the parts of 10 and 2 and 8.[40]

I-23*.   IF ONE OF THE PARTS OF A GIVEN NUMBER IS DIVIDED BY A KNOWN
NUMBER AND THIS QUOTIENT IS DIVIDED BY THE OTHER PART SO THAT THE
FINAL QUOTIENT IS KNOWN, THE TWO PARTS CAN BE FOUND.[41]

Divide x by c the given number to obtain z, which in turn is divided by
y to obtain e. Now y times e equals z and z times c is x. Hence, if c is

multiplied into e to obtain f, then f is known. Moreover, y times f is x. So x and y are found.

For example, divide 10 into two parts and divide one of them by 5. Divide the quotient by the other part so that the new quotient is a fifth. This times 5 is one, which is a portion of one of the parts multiplied by the other. Add one to this to get 2, which divided into 10 yields 5 which was one of the divisors.

I-24.   IF THE SUM OF AN ALIQUOT PART OF THE QUOTIENT OF THE TWO PARTS
OF A GIVEN NUMBER AND ONE OF THE PARTS IS KNOWN, THEN THE TWO PARTS
CAN BE FOUND.[42]

Let c be the quotient of x and y and its half[43] be d, so that $x + d$ is known. You must now consider which is larger, $x + y$ or $x + d$, both given.

Let the larger be $x + y$. To this add just the aliquot part of c, namely e equal to $\frac{1}{2}$. Now, since twice d times y equals x and twice d times e is d itself, it follows that d times twice $e + y$ is $x + d$. Set g equal to the difference of $x + y + \frac{1}{2}$ and $x + d$. Note that twice the product of d and $d + g$ is $x + d$; hence, half the one equals half the other. Since this is known as is g, d and x can be found.[44]

Consider now that $x + d$ is the larger. Then y times $y + g$ is half of $x + y$. In a similar fashion their difference can be found.

Finally, if $x + d$ and $x + y$ are equal, then the sum of y squared and half y is half of $x + d$. The solution follows as in the second case.

For example, divide the two parts of 10 by one another and take their half. Add this to the dividend so that the sum equals $4\frac{1}{3}$. Subtract this from $10\frac{1}{2}$ to get $6\frac{1}{6}$. Take half of $4\frac{1}{3}$, quadruple it as is customary, and add the square of $6\frac{1}{6}$ to get $46 + \frac{2}{3} + \frac{1}{36}$(*sic*), whose root is $6 + \frac{2}{3} + \frac{1}{6}$. From this subtract $6\frac{1}{6}$ to get $\frac{2}{3}$, whose half is $\frac{1}{3}$. Subtract this in turn from $4\frac{1}{3}$ to obtain 4, the second of the two parts.

I-25.   IF THE PRODUCT OF A KNOWN NUMBER AND THE SUM GIVEN ABOVE
IS KNOWN, THEN BOTH PARTS CAN BE FOUND.

Let d be the product of x and the given c, and let e be the quotient of this product and y. Call f the aliquot part of e. Now the sum of d and f is given. Divide all of this by c to produce $g + h$. Let g equal x; and f is the product of c and h. Since c is to y as e is to x and because c times h is

f, it follows that y times h is the aliquot part of x. Because $g + h$ is known, so is $x + h$. Therefore x and y can be found.[45]

For example, let 5 times one part of 10 together with half the quotient of this product and the other part equal 50. Divide this by 5 to get 10. Now proceed as in the third case of theorem 24. Quadrupling half of 10 gives 20 to which is added the square of $\frac{1}{2}$, namely $\frac{1}{4}$, to obtain $20\frac{1}{4}$. The root of this is $4\frac{1}{2}$. From this subtract $\frac{1}{2}$ and halve the remainder to get 2, which is one of the parts.

I-26. IF THE SUM OF THE TWO QUOTIENTS FORMED BY DIVIDING THE TWO PARTS OF A GIVEN NUMBER BY TWO DIFFERENT KNOWN NUMBERS IS GIVEN, THEN THE PARTS CAN BE FOUND.

Let c divide x and d divide y to form the given sum $e + f$, and c and d are given. Assume c to be greater than d and let their difference be g. Set the product of d and $e + f$ equal to $n + m$ with m equal to y. Let the difference of $x + y$ and $m + n$ be j. Divide j by g to get e. Therefore x and y can be found.[46]

For example, separate 10 in the usual manner and divide the first part by 3 and the other by 2. Set their sum equal to 4. Multiply this by 2 to reach 8, which subtracted from 10 leaves 2. Divide this by 1, the difference of 3 and 2. The quotient is 2, which multiplied by 3 yields 6, one of the parts of the given number.

I-27. IF THE PRODUCT OF THE QUOTIENTS IN THEOREM 26 IS KNOWN, THEN THE TWO PARTS CAN BE FOUND.

Multiply e by f to produce the known b. Let the product of c and b be h, which is the same as multiplying f by the product of c and e, or by x. Then multiply d by h to produce j, which is the same as multiplying x by the product of d and f, or by y. Since this product has been found, the two parts can be found.

For example, divide one part of 10 by 4 and the other by 2 and set the product of the quotients equal to 2. Multiply this 2 by 4 and the product by 2 to get 16, which equals the product of the two parts.[47] Hence, it is evident that the two parts are 8 and 2.[48]

I-28α. IF THE QUOTIENT OF THE TWO PARTS OF A GIVEN NUMBER IS KNOWN, THEN THE TWO PARTS CAN BE FOUND.

Let the given quotient[49] of e and f be h, and let the quotient of h and d equal k, and set the product of k and c equal to j. Now since the product of f and h is e, so is the product of y and k equal to e.[50] Thus the product of y and j is x. Since the quotient of x and y is now known, x and y can be found.[51]

For example, separate 10 into two parts and let a fourth of one be divided by half the other to obtain $\frac{1}{3}$. Quadruple a half of this to obtain $\frac{2}{3}$.[52] Divide therefore in the customary way 10 by $1\frac{2}{3}$ to obtain 6. This is one of the parts of 10.

I-28β.   IF A NUMBER IS DIVIDED INTO TWO PARTS AND EACH DIVIDES THE SAME GIVEN NUMBER SO THAT THE SUM OF THE QUOTIENTS IS KNOWN, THE PARTS CAN BE FOUND.[53]

Divide c by a and b so that the sum $d + e$ is known. Then divide c by $a + b$ to obtain f. Since the product of f and the square of $a + b$ (call it g) equals the product of $d + e$ and h (which is the product of a and b), it follows that the product of f and g equals the product of c and $a + b$. So multiply $a + b$ by e and divide the product by $d + e$, and this will produce h. Hence, both a and b will be found.

For example, let each of the two parts of 10 divide 40 so that the sum is 25. Multiply 40 by 10 and divide the product by 25 to obtain 16. And that is the product of the two parts.

I-29.   IF THE PRODUCT OF ONE PART OF A GIVEN NUMBER AND THE GIVEN NUMBER IS EQUAL TO THE SQUARE OF THE OTHER PART, THEN EACH PART CAN BE APPROXIMATED.[54]

Let the product of $x + y$ and y equal x squared. Now, since the square of $x + y$ equals the sum of the products of $x + y$ times x and of $x + y$ times y, it also equals the sum of x squared and x times $x + y$.[55] Now since $x + y$ was given, both x and y can be found.[56]

For example, separate 10 into two parts so that 10 times one part equals the square of the other part. Now 10 squared is 100. Quadrupling this yields 400. To this is added, as is customary, the square of 10 to make 500. Take the closest root of this, which is $22\frac{1}{3}$. From this subtract 10. Half of this remainder is $6\frac{1}{6}$. And this is the greater part[57] which was squared.

End of Book I

Book Two

II-1. IF THERE ARE FOUR NUMBERS IN PROPORTION AND THREE OF THEM
ARE GIVEN, THEN THE FOURTH CAN BE FOUND.
By cross multiplication the products are equal. Taking them alternately
then, multiply the two given numbers and divide by the third to produce
the number not given at first.

For example, let 20 be to some number as 5 is to 4. Now by multiplying
the given antecedent by the known consequent of the other, that is 20
times 4, one obtains 80. Divide this by 5 to get 16, which is the ungiven
consequent of the aforementioned 20.

II-2. IF THE RATIO OF A KNOWN NUMBER TO SOME OTHER NUMBER IS
GIVEN, THEN IT CAN BE FOUND.[58]
It is just as easy in multiple proportion as in partial proportion to solve
this as long as the consequent is given. And this is hardly an absurd
remark. Simply multiply it by the value of the given ratio, adding the
fractions where necessary, and you have the antecedent. If the antecedent
were given, then divide it by the denomination and that is the consequent.
To put it another way: add the fractions and multiply by the consequent,
and you have the consequent. This completes the mechanics.

For example, let the ratio of 100 to some number be equal to $3 + \frac{1}{2} +
\frac{1}{4}$.[59] Divide 100 by $3\frac{3}{4}$. This gives $26\frac{2}{3}$, which is the consequent. Again,
let the ratio of $26\frac{2}{3}$ to some number be $\frac{1}{4} + \frac{1}{60}$. The sum of these is $\frac{16}{60}$.
Multiply 60 by $26\frac{2}{3}$ to get 1600 and divide then by 16 to get 100. This is
the consequent.

II-3. IF THE RATIO OF THE FIRST TO THE SECOND IS GIVEN, THEN SO IS THAT OF THE SECOND TO THE FIRST.

Divide one by the denomination of the ratio of the first to the second, and what results is the denomination of the second to the first.

For example, let the ratio of the first to the second be $2\frac{2}{3}$. So divide one by $2\frac{2}{3}$ to get $\frac{3}{8}$. Therefore the second ratio is $\frac{3}{8}$ of the first.

II-4. IF THE RATIO OF A WHOLE NUMBER TO ONE OF ITS PARTS IS GIVEN, THEN THE RATIO OF THE OTHER PART TO THE FIRST CAN BE FOUND: AND IF THE RATIO OF THE OTHER TO THE FIRST IS KNOWN. THEN SIMILARLY IS FOUND THE RATIO OF THE WHOLE TO THE FIRST PART.

This is simple. Let one be subtracted from the ratio of the whole to the first part. What remains is the ratio of the other part to the first. By the same token, if one is added to the ratio of the two parts, the ratio of the whole to the parts results.

For example, let the ratio of 10 to 3 be $3\frac{1}{3}$. Therefore the ratio of 7 to 3 is $2\frac{1}{3}$. Conversely, the ratio of 7 to 3 is $2\frac{1}{3}$; whence the ratio of 10 to 3 is $3\frac{1}{3}$.

II-5. IF THE RATIO OF A KNOWN NUMBER TO A PART IS GIVEN, THEN THE RATIO OF THE WHOLE TO THE OTHER PART CAN BE FOUND.[60]

If the ratio of the whole to a part is given, then the ratio of the other part to the first can be found. Whence the ratio of the first to the other is known; therefore is found the ratio of the whole to the other part.

For example, let 6 be one part of 10 and let the ratio of 10 to 6 be $\frac{5}{3}$. Therefore the ratio of 6 to 4 is $\frac{2}{3}$. Now divide one by $\frac{2}{3}$ and the result is $\frac{3}{2}$; whence the ratio of 6 to 4 is $\frac{3}{2}$. And the ratio of 10 to 4 is $\frac{5}{2}$.

II-6. IF THE RATIO OF THE TWO PARTS OF A GIVEN NUMBER IS KNOWN, THEN EACH OF THEM CAN BE FOUND.

If the ratio of the two are known, then the ratio of the whole to the same part is known. Now since the whole is known, so will that part be found and consequently the other part also.

For example, separate 10 into two parts of which one is the quadruple of the other. Therefore 10 will be the quintuple of it. So the other part is 2.

II-7.   IF THE RATIO OF A NUMBER TO A SECOND IS GIVEN TOGETHER WITH THE RATIO OF THE SECOND TO A THIRD, THEN THE RATIO OF THE FIRST TO THE THIRD CAN BE FOUND; AND CONVERSELY.[61]

Let the denomination of the ratio of the first to the second be multiplied by the denomination of the ratio of the second to the third; and this becomes the ratio of the first to the third. Likewise, let the ratio of the first to the third be divided by the ratio of the second to the third. What results is the ratio of the first to the second.

For example, let the ratio of the first and the second be $\frac{10}{7}$ and that of the second and the third be $\frac{7}{5}$. Multiplying these together, the ratio of the first and the third is found to be 2. Hence the first number is twice the third. Again, dividing 2 by $\frac{7}{5}$ produces $\frac{10}{7}$. Hence the first is $\frac{10}{7}$ of the second number.

II-8.   IF HOWEVER SO MANY NUMBERS HAVE EACH A GIVEN RATIO TO A DEFINITE NUMBER, THEN THE RATIO OF ALL OF THEM TO THE SAME NUMBER CAN BE FOUND.

Let the denominations of all the ratios to the definite number be added and the sum will be the denomination of the whole to the same number.

For example, let the individual ratios of three numbers to 4 be $\frac{4}{3}, \frac{9}{4}$, and $\frac{5}{2}$. Adding the ratios produces the ratio of their sum to 4, namely, $\frac{13}{32}$.

II-9.   IF THE RATIOS OF A DEFINITE NUMBER TO HOWEVER SO MANY NUMBERS ARE GIVEN, THEN THE SUM OF ALL SUCH RATIOS CAN BE FOUND.

As the one to all of those is known, so is the ratio of all of them to it found. Hence, the sum of them to it can be found as was desired.

For example, let the individual ratios of one to two other numbers be $\frac{5}{3}$ and $\frac{3}{2}$. Divide each of these into 1 to obtain $\frac{3}{5}$ and $\frac{2}{3}$. The sum of these is $\frac{19}{15}$.[62] Divide this into one to obtain the sum of the ratios, namely, $\frac{15}{19}$.

II-10.   IF FROM TWO GIVEN NUMBERS TWO PARTS ARE SUBTRACTED AND THE RATIOS OF THE SUBTRAHENDS AND THE REMAINDERS ARE KNOWN ALTHOUGH UNEQUAL TO THE RATIO OF THE GIVEN NUMBERS, THEN EACH PART CAN BE FOUND.

Let the given numbers be $x + y$ and $z + w$, from which x and z are subtracted. Since the ratio of x and z does not equal the ratio of the given numbers, it will also not equal the ratio of y and w. Now let x be to e as y is to w. Therefore $x + y$ is to $e + w$ as y is to w. Since $x + y$ was given, $e + w$ is found. Let g be the difference of z and e. Since the ratios of z to x and of e to x are known, that of g to x can be found. Since g is known, x can be found. Proceed similarly for the others.

For example, let the given numbers be 20 and 12, the ratio of the subtrahends be 2, and of the remainders be $\frac{3}{2}$. Now the ratio of 20 to some number, namely 10, is 2, and the difference of 12 and 10 is 2. And knowing the ratios of the subtrahends and of the remainders, it follows that the subtrahend for 20 is 8 and its remainder is 12. Finally, the subtrahend of 12 is 4 and the remainder is 8.

II-11. IF THE RATIO OF TWO NUMBERS IS GIVEN AND A KNOWN NUMBER IS SUBTRACTED FROM ONE OF THESE AND ANOTHER GIVEN NUMBER ADDED TO THE OTHER SO THAT THE SUM AND DIFFERENCE FORM A GIVEN RATIO, THEN THE PARTS CAN BE FOUND.

Let the given ratio be of $a + x$ to y. And let the known a be subtracted from $a + x$. Then add the known d to y, so that the ratio of x to $y + d$ is known. Let this be equal to the ratio of e to a, which is equal to the ratio of $y + d$ to x. Hence $y + d + e$ to $x + a$ is a known ratio. But y to $a + x$ is a given ratio. Therefore the ratio of $d + e$ to $x + a$ is known. Since $d + e$ is known, then $a + x$ and y are found.

For example, let the ratio of the larger to the smaller be $1\frac{1}{3}$. Subtracting 7 from the larger and adding 6 to the other, then the whole of the smaller is twice what remains of the larger. Let the number then be twice 7, namely 14, which is added to the whole of the smaller. Then the sum becomes twice the larger and increases the lesser by 20. Because the lesser is $\frac{3}{4}$ of the greater, take away $\frac{3}{4}$ from 2 so that $\frac{5}{4}$ remains. Hence 20 contains the larger once and a fourth to make it 16, while the smaller is 12.

II-12α. IF GIVEN NUMBERS ARE ALTERNATELY ADDED TO AND SUBTRACTED FROM TWO NUMBERS AND THE RATIOS OF THE SUM AND DIFFERENCE ARE KNOWN, THEN BOTH THE NUMBERS CAN BE FOUND.

Let the numbers be $a + x$ and $d + y$. Let a, d, c, and f be given, as are the ratios of $a + x + c$ to y and of $d + y + f$ to x. Hence $a + x$ and $d + y$ are known. Since $a + x + c$ to y is given, subtract from it the number $a + c$ and add $d + f$ to the other. $d + y + f$ to x is found by the previous operation.[63]

For example, let 4 be subtracted from the smaller and 2 added to the other so that the ratio of the two is 2. Now if 3 is added to the smaller and 4 subtracted from the greater, the sum is $\frac{4}{3}$ of the remainder. By the previous operation, therefore, the greater number is 16 and the lesser is 13.

II-12β.   IF TWO GIVEN NUMBERS ARE SUBTRACTED FROM TWO NUMBERS IN A KNOWN RATIO SO THAT THE REMAINDERS ARE IN A GIVEN RATIO, THEN THE WHOLE NUMBERS CAN BE FOUND.

Given the ratio of $x + a$ and $y + b$ with a and b known together with the ratio of x and y, let e be to a as y is to x. Also let the difference of e and b be g. Therefore the ratio of $e + y$ to $x + a$ is known as is the ratio to $y + b$. Also is found the ratio of g to $y + b$. Since g is known, so is the ratio of $y + b$ to $x + a$ found.[64]

For example,[65] let the ratio of two numbers be two. From the larger subtract 4 and from the smaller take 6 so that the ratio of the remainders is 4. Quadrupling 6 yields 24, which added to the remainder of the larger gives a whole that is four times the smaller and twice the larger. This is the difference, 20, which is equal to the larger number. The smaller is 10.

II-13α.   IF FROM TWO GIVEN NUMBERS TWO OTHER NUMBERS WHOSE RATIO IS KNOWN ARE SUBTRACTED SO THAT THE DIFFERENCE OF THE REMAINDERS IS KNOWN, THEN EACH OF THE NUMBERS CAN BE FOUND.

Let the following be given: $w + x$, $z + y$, the ratio of x and y, e the difference of w and z, g the difference of x and y, and f the difference of $w + x$ and $y + z$. Now if w is greater than z, regardless of whether e or f is larger than the other, then their difference will be the same as that of x and y. But if w is less than z, then the sum of e and f equals g.

For example, let the given numbers be 15 and 9. Let what is subtracted from 15 be thrice that taken from 9. Let 2 equal the difference of the remainder of 15 and the remainder of 9, which added to the difference of 15 and 9 makes 8. Therefore these numbers are 12 and 4, and the remainders are 3 and 5.

II-13β. TWO NUMBERS ARE GIVEN IN RELATION TO ONE ANOTHER. IF A GIVEN NUMBER IS SUBTRACTED FROM ONE OF THESE AND FROM THE OTHER IS TAKEN [A FRACTIONAL PART OF IT] SO THAT THE RATIO OF THE REMAINDERS IS KNOWN, THEN THE TWO WHOLE NUMBERS CAN BE FOUND.
For if one will be given with respect to its subtrahend, it will be known with respect to its remainder. Hence similarly with respect to the remainder of the other. Because the like is true with respect to the whole, similarly is it with respect to the subtrahend. Hence the two numbers are found.

For example, let the ratio of two numbers be $1\frac{1}{2}$. From the greater subtract 4, from the other a fourth and an eighth of itself, so that the ratio of the remainders is 2. Since the subtrahend of the smaller number is $\frac{3}{8}$ of it, its remainder will be $\frac{5}{8}$ of itself, which is half the remainder of the larger number. Then the remainder will be the entire smaller number and its fourth. The subtrahend of the greater, that is, 4, will be its other fourth. Hence it will be 16 and the other will be 24.

II-14α. IF TWO NUMBERS WHOSE RATIO IS KNOWN ARE SUBTRACTED FROM TWO GIVEN NUMBERS AND THE PRODUCT OF THE REMAINDERS IS ALSO KNOWN, THEN EACH OF THE NUMBERS CAN BE FOUND.
Let the given numbers be $x + y$ and $z + w$, the given product yw equal to f, and the ratio of x to z be known as d. Let the ratio of x to z equal the ratio of $x + y + c$ to $z + w$; hence the ratio of $y + c$ to w is found. Therefore the product of y and $y + c$ to f, and the difference of $y + c$ and y, are known. Hence both y and $y + c$ are found;[66] and the rest can be found.

For example, let the given numbers be 12 and 10, and let the ratio of the numbers subtracted from 12 and 10 be $1\frac{1}{2}$, and the product of the remainders be 36. Now since the ratio of 15 to 10 is $1\frac{1}{2}$ as is the ratio of 54 to 36, that which is contained in the two numbers 15 and 12 whose difference is 3 will be 9, and the other is 6 which is one part of 12, the other being 6. Consequently the parts of 10 are 4 and 6.

II-14β. TWO NUMBERS ARE GIVEN IN RELATION TO ONE ANOTHER. IF A GIVEN NUMBER IS SUBTRACTED FROM ONE AND A FRACTIONAL PART OF

THIS REMAINDER IS ADDED TO THE OTHER TO FORM A GIVEN SUM; THEN EACH OF THEM CAN BE FOUND.

Let the two numbers be $a + b$ and c, let a be given, let a fourth of b be added to c to make the entire $c + d$ known. Now quadruple $c + d$ and get $e + f$, with f the quadruple of c and e equal to b. Now add h the equal of a to $e + f$. Hence $h + e$ equals $a + b$. And because the ratio of $a + b$ to c is known, so is the ratio of $h + e$ to its quadruple, namely f. Therefore $h + e + f$ to $h + e$ is a known ratio. Since $h + e + f$ is known, so are $h + e$ and $a + b$ found.

For example, if the ratio of one to another is $1\frac{1}{4}$, subtract 4 from the one and add half the remainder to the other to make 11. Double 11 to get 22 and add to it 4 to obtain 26. Since the double of the subsesqui-octave[67] of the three contains the other and its three fifths, so will the 26 contain it twice and its three fifths. Therefore it is 10 and the other is 8.

II-15.   IF A NUMBER IS SEPARATED INTO HOWEVER SO MANY PARTS OF WHICH ONE IS GIVEN AND THE RATIOS OF THE WHOLE TO EACH OF ITS RE-MAINING PARTS ARE GIVEN, THEN THE WHOLE NUMBER CAN BE FOUND.

If the ratios of the whole to each of the remaining parts are known, then its ratio to the sum of the parts is found as well as to the given part. Hence the whole number can be found.

For example, separate the whole number into four parts such that the first is $\frac{1}{3}$ of it, the second is $\frac{1}{4}$ of it, the third is $\frac{1}{5}$ of it, and the fourth equals $6\frac{1}{2}$. Now $\frac{1}{3}$, $\frac{1}{4}$, and $\frac{1}{5}$ equal $\frac{47}{60}$. Hence $6\frac{1}{2}$ is $\frac{13}{60}$ of the whole number, which must therefore be 30. Dividing 30 into its parts produces 10, $7\frac{1}{2}$, and 6, leaving $6\frac{1}{2}$ for the last part.

II-16.   IF A NUMBER IS SEPARATED INTO HOWEVER SO MANY PARTS SUCH THAT ONE OF ITS PARTS ADDED TO A GIVEN NUMBER FORMS A KNOWN RATIO WITH THE WHOLE, AND THE RATIOS OF THE OTHERS PARTS TO THE FIRST ARE KNOWN, THEN THE WHOLE NUMBER CAN BE FOUND.

Since each of the parts is in a known ratio to the whole and the sum of these can be found, it follows that the given number that was added to a part has a discoverable ratio to the whole number. Therefore the whole number can be found.

For example, separate a number into three parts such that the first is $\frac{1}{3}$ of it, the second is $\frac{1}{4}$ of it, and the third increased by 3 is $\frac{2}{3}$ of the whole. The sum of all these makes $\frac{3}{4}$ of the whole. Consequently $\frac{1}{4}$ of the whole is 3, or the whole is 12.

II-17. IF A NUMBER IS SEPARATED INTO HOWEVER SO MANY PARTS SUCH THAT ANY ONE OF THE PARTS WITH SOME GIVEN NUMBER IN A KNOWN RATIO WITH THE WHOLE AND THE RATIOS OF THE REMAINING PARTS TO THE WHOLE ARE KNOWN, THEN THE WHOLE CAN BE FOUND.[68]
Since each of the remaining parts has a known ratio with the given number, the sum of these ratios can be found. Likewise there is a ratio between what remains and the whole number, which can be added to the previous sum. Since this sum is known, the whole number can be found.

For example, separate the number as before into three parts of which the first is $\frac{1}{2}$ of it, the second is $\frac{1}{3}$ of it, and the third with $\frac{1}{4}$ of it equals 5. Since $\frac{1}{2}$ and $\frac{1}{3}$ are $\frac{5}{6}$, what remains is $\frac{1}{6}$ of the whole. By adding $\frac{1}{4}$ of the whole to this, 5 is had. In other words, 5 is $\frac{5}{12}$ of the whole which is 12.

II-18. IF A GIVEN NUMBER IS DIVIDED INTO HOWEVER SO MANY PARTS WHICH ARE IN KNOWN CONTINUED RATIOS, THEN ANY OF THE PARTS CAN BE FOUND.
Since the ratio of the first part to the second and of the second to the third part are given, so also is known the ratio of the first to the third. Similarly is formed the ratio of the first to the sum of the other two parts. Hence the ratio of the whole to the first part is known. Therefore the first part, as well as the other two parts, can be found.

For example, separate 60 into three parts, of which the greatest is twice the second and the second is thrice the third. Hence the first is six times the third. Therefore the smaller parts are $\frac{2}{3}$ of the greatest. Hence the whole number is $1\frac{2}{3}$ of its greatest part. Consequently the parts of 60 are 36, 18, and 6.

II-19. IF TO EACH OF HOWEVER SO MANY PARTS OF A GIVEN NUMBER DEFINITE NUMBERS ARE ADDED AND THE SUMS ARE IN KNOWN CONTINUED RATIOS, THEN AS BEFORE[69] THE RATIO OF THE SUM OF THE PARTS CAN BE FOUND AS WELL AS EACH PART.[70]
Let the definite numbers be added to the several parts of a given number, and let the continued ratios of these sums be known. Hence, the sum of these ratios can be found. If the definite numbers are subtracted from the individual sums, then the parts of the given number can be found.

For example, divide 20 into three parts. Add 4 to the first, 1 to the second, and 5 to the third. Let the ratio of the first two be $\frac{3}{2}$ and of the second two be 2. By adding the addends to 20 one obtains 30. And if 30 is separated according to the ratios, the parts are 15, 10, and 5. So take the addends severally from these, namely 4 from 15, 5 from 10, and 1 from 5. The respective remainders are 11, 5, and 4. These total 20.

II-20.   IF CONSTANTS ARE ADDED TO EACH PART OF A NUMBER AND THE SUMS SET IN GIVEN RATIOS TO EACH SUCCEEDING PART SO THAT THE LAST SUM IS IN A GIVEN RATIO TO THE FIRST PART, THEN ALL THE PARTS CAN BE FOUND.

To the three parts of a let the numbers d, e, and f be added to form $x + d$, $y + e$, and $z + f$. Let the ratios of $x + d$ to y, $y + e$ to z, and $z + f$ to x be given. Now let $x + d$ to y be as g to e, and this is known. Therefore $g + d + x$ to $y + e$ is the same as $x + d$ to y. Similarly, since $y + e$ to z is known, so also is $g + d + x$ to z. Likewise let h to f be as $g + d + x$ to z, and this makes $h + d + g + x$ to $z + f$ known. But since $z + f$ to x is known, then is known $h + g + d + x$ to x as is $h + g + d$ to x. But since $h + g + d$ is known, so is x found as are y and z.

For instance, let there be three numbers such that the ratio of the first increased by 6 to the second is $\frac{5}{3}$, the second increased by 4 to the third is 2, and the third increased by 2 to the first is $\frac{5}{7}$. Add $6\frac{2}{3}$ to the first, which now becomes $12\frac{2}{3}$ whose ratio to the second is twice the third. Hence the ratio of the first to the third is $\frac{10}{3}$ because $\frac{20}{3}$ is twice $\frac{10}{3}$. Now the third increased by 2 is $\frac{5}{7}$ of the first. Hence the first plus $10\frac{1}{3}$ is $2 + \frac{2}{7} + \frac{2}{21}$ of itself, which is to say $19\frac{1}{3}$ is $1 + \frac{2}{7} + \frac{2}{21}$ of the first. Therefore, the first part is 14, the second is 12, and the third is 8.

II-21.   IF ONE PART OF A GIVEN NUMBER (OR EVEN BOTH PARTS)[71] HAS A KNOWN RATIO TO A CONSTANT SO THAT A NEW SUM WITH THE OTHER PART IS GIVEN, THEN BOTH PARTS CAN BE FOUND.[72]

Divide $x + y$ into x and y. Let the ratio of c to y be given and also $x + c$. Let the difference of $x + c$ and $x + y$ be known as e, which is equivalent to the difference of c and y. Since the ratio of c and y is known, likewise is known the ratio of e and y. But since e is known, y and x can be found.

For instance, separate 12 into two parts so that one part plus $\frac{1}{3}$ of the second is 6. Let the difference of this and 12 be 6, which is also the difference of the $\frac{1}{3}$ and its whole. Since 6 is $\frac{2}{3}$ of the whole or 9, what remains is 3 or $\frac{1}{3}$ of 9. This with 3 makes 6.[73]

II-22.   IF ONE PART OF A GIVEN NUMBER INCREASED BY A CONSTANT (OR EVEN A MULTIPLE OF IT WITH A CONSTANT) HAS A KNOWN RATIO TO THE OTHER PART, THEN BOTH PARTS CAN BE FOUND.

As before separate the known $x + y$ into x and y, and let $x + c$ be in a known ratio to y. Then add c to $x + y$ to become $x + y + c$, which in turn can be separated into y and $x + c$,[74] which was the given ratio. Therefore, both parts can be found.

For example, separate 12 into two parts so that the first increased by 2 is $\frac{3}{4}$ of the other. Add 2 to 12 to get 14, which is to be separated in the given ratio, namely into 8 and 6. Subtract 2 from the latter to obtain 4.

II-23.   IF A GIVEN NUMBER IS DIVIDED INTO SEVERAL PARTS SO THAT EACH IS IN A GIVEN RATIO TO THE SUM OF THE NEXT TWO PARTS AND THE NEXT TO THE LAST HAS A KNOWN RATIO WITH THE SUM OF THE LAST AND THE FIRST PART, THEN ALL THE PARTS CAN BE FOUND.

Separate a given number into four parts so that the three ratios are known, namely w to $x + y$, x to $y + z$, and y to $z + x$. Divide w into e and f, and let the ratios of x to e and y to f be as $x + y$ to w. Likewise, since the ratio of x to $y + z$ is known, as is the ratio of x to e, the ratio of e to $y + z$ has been found; and this can be divided into g and h according to the ratio of y to z. Since $f + g$ to y is known, as is y to $w + z$, it follows that $f + g$ to $z + w$ is found. Separate g into k and j according to the ratio of z to w so that w equals $h + k + j$. Since j to w is found, similarly is $h + k$. But the ratio of $h + k$ to z is known, whence is the ratio of z to w. The ratio of h to $w + x$ is found, whence the ratio of w and y are found. It is the same with $w + z$. Therefore all of them can be found with respect to themselves and to the whole. So the individual parts are found.

Should there be five parts to the given number, such as u, w, x, y, and z, then divide u into $y + z$. For the same reason divide z into $y + x$ and y into $w + x$ and x into $w + u$. Then the ratio of u to w is found. Proceed similarly to find the others.

For example, separate 32 into four parts such that the first is $\frac{1}{7}$ of the second and third, the second is $\frac{1}{5}$ of the third and fourth, and the third is $\frac{1}{2}$ of the fourth and first. Then multiply the second and third ratios by

$\frac{1}{7}$ and the third again by $\frac{1}{5}$. By appropriate substitution from the second and third ratios into the first, one obtains the result that the first part is $\frac{1}{8}$ of the fourth. By similar substitutions, one realizes that the second part is $\frac{5}{16}$ of the fourth, the third is $\frac{9}{16}$ of the fourth, and the fourth is 16 itself. Hence the other parts are 2, 5, and 9.

II-24.   IF THE SUM IS TAKEN OF THE PARTS OF A NUMBER, TO EACH OF WHICH A GIVEN NUMBER IS ADDED SO THAT THIS SUM IS IN A GIVEN RATIO TO THE SUM OF THE REMAINING PARTS OF THE NUMBER, THEN EACH OF THE PARTS CAN BE FOUND.

Let the four parts of the number be w, x, y, and z, and the given addend be e. Since the ratio of $w + e$ to $x + y + z$ is given, separate this in such a fashion that the ratios of f to x, g to y, and h to z are known even as is $w + e$ to $x + y + z$. Likewise let k to e be as f to x. Because $x + e$ to $y + z + w$ is known, as is $f + k$ to $x + e$, then is known $f + k$ to $y + z + w$. Divide again according to their ratios into l, m, and n and let the ratio of y to w equal g to y. Hence, $k + e + w + y$ equals the product of $y + n + l + g$ and h. But $x + e$ to $w + y + z$ is known, and $w + y + z$ to $n + y + l + g$ times h equals $k + e + w + y$. Hence $x + e$ to $k + e + w + y$ is known. But with the ratio of $k + e + w + y$ and $e + w + m$ known, $w + y + z$ is found. Hence, $y + e$ to $w + e + m$ is known, similarly $w + e + m$ to $y + e$ and to $z + e$. And so $w + e$ to thrice e and $x + y + z$ is found. But $w + e$ to $y + z + x$ is known. Hence thrice e to $x + y + z$ is found. So is e found, then w, and so on.

For example, let the given number be 6, and let there be four numbers. The first plus 6 is a ninth of the remaining; the second plus 6 is a third of the other three; the third plus 6 is three fifths of the others; the fourth plus 6 is equal to the sum of the other three. Multiply the third by the ninth to obtain a twenty-seventh. Likewise multiply the third by $1\frac{1}{5}$ to obtain $\frac{1}{3}$ and $\frac{1}{27}$, which are to be divided by $\frac{1}{9}$ and $\frac{1}{27}$, whence arise $2\frac{1}{2}$. Therefore the second and 6 contains the first and 6, $7\frac{1}{2}$ times. Likewise multiply the ninth by $\frac{3}{5}$ to obtain $\frac{1}{15}$, and $\frac{3}{5}$ by $1\frac{1}{9}$ to get $\frac{2}{3}$ and $\frac{1}{15}$, which divided by $\frac{1}{19}$ and $\frac{1}{15}$ will produce $3\frac{3}{4}$. But the third plus 6 contains the first plus 6, $3\frac{3}{4}$ times. Since the fourth plus 6 are as the rest multiplied by one, so $\frac{1}{9}$ times one is $\frac{1}{9}$. But if one is multiplied by $1\frac{1}{9}$, the result is $1\frac{1}{9}$, which divided by $\frac{2}{9}$ yields 5. Wherefore the fourth plus 6 is 5 times the first plus 6. And so the second, third, fourth, and thrice 6 contains the first plus 6, $11\frac{1}{4}$ times. But those three contain it nine times. Wherefore thrice 6 contains the first plus 6, $2\frac{1}{4}$ times. Thus they are 8. From this subtract 2, which is the first. Now the second plus 6 equals 20, whence the second unknown is now 14. The third plus 6 equals 30 and so it is 24. The fourth plus 6 is 40 and so it is 34.[75]

II-25. IF A GROUP OF NUMBERS IS PROPOSED SUCH THAT EACH WITH SOME MULTIPLE OF ITS SUCCESSOR EQUALS THE SAME GIVEN NUMBER, THEN EACH OF THE GROUP CAN BE FOUND.[76]

Let the given number be e and the others these four, w, x, y, and z. Let the known f be the multiple of w, similarly g for x, h for y, and k for z. Then each of $w + g$, $x + h$, $y + k$, and $z + f$, all equal e. Now as g is to x, so is j to h and j to y. Likewise m is to k, and as m is to z so is n to x. Therefore $g + j$ and $j + m$ and $m + n$ are known.[77] Taking the difference of $w + g$ and $g + j$, one obtains the difference of w and j. Similarly with j and n, the difference of w and n result. But since the difference of a and n is known, both of them can be found. Then subtract w from e to obtain g. And so on. Now assume x, and five other numbers, g, j, m, n, and t. Since the difference of x and j is known, as is j and n, then the difference of a and n is found. And since $n + y$ is known, if a is less than n and given as such, subtracting this from $n + y$ produces $w + y$ as known. And so each is found.

For example, let the given number be 119. There are four other numbers. The first with half the second, the second with a third of the third, the third with a fourth of the fourth, and the fourth with a fifth of the first: each sum equals 119. Multiply the second sum by $\frac{1}{2}$, the third by $\frac{1}{6}$, and the fourth by $\frac{1}{24}$. Subtracting half the sum of the second from the first yields $59\frac{1}{2}$, which equals the difference of the first number and $\frac{1}{6}$ the third. Subtracting $\frac{1}{24}$ the fourth sum from $\frac{1}{6}$ the third sum yields $14\frac{21}{24}$. Adding these two results in $74\frac{9}{24}$, the difference of the first number and $\frac{1}{120}$ of it. Multiplying this by 120 yields 8,925.[78] Now the first number is to that as 120 is to 119. Hence if the product is divided by 119, the first number results, namely, 75. Subtracting this from 119 and doubling the remainder yields the second number, 88. Subtract this from 119 and triple the remainder to obtain the third number, 93. Subtract this from 119 and quadruple the result to obtain the fourth number, 104. Subtract this from 119 and quintuple the remainder to obtain once again the first number, 75.

II-26.   FOR ANY NUMBER OF NUMBERS: IF THE SUM OF ONE OF THEM AND
THE GIVEN RATIO OF A KNOWN NUMBER AND THE SUM OF THE REST OF THEM
EQUALS A GIVEN NUMBER, THEN EACH OF THEM CAN BE FOUND.

Let the given number be z and the proposed numbers a, b, c, d. Let the
ratio of $e + f + g$ to $a + b + c$ be given, which with d equals z. Similarly,
the ratio of $h + k + l$ to $b + c + d$ is given, which with a equals z; the
ratio of $m + n + o$ to $c + d + a$ is given, which with b equals z; and
finally, the ratio of $p + q + r$ to $d + a + b$ is given, which with c equals z.
Let the ratio of $h + k + l$ to $b + c + d$ be less than the ratio of $m + n + o$
to $c + d + a$, both of which are less than the ratio of $p + q + r$ to $d + a + b$,
and all of these are less than the ratio of $e + f + g$ to $a + b + c$. Now
since k and l are less than n and o, let them be subtracted and call the
remainder $f + c$. Also, $a + h$ is larger than $m + b$; but if h is greater than
b, then a is less than m because the ratio of m to a is larger than that of h
to b. Now if a is subtracted from m and b from h, what remains is $x + v$;
and v will be as x, s, t. Because $m + o$ is less than $p + r$ and b less than q,
because it is less than h, c will be less than n. Because $p + a$ is less than
$e + f$ and c is less than g, since it is less than n, d will be less than r. Now
therefore, since v is resolved into $x + s + t$ which is given with respect to
$a + c + d$, similarly is b resolved into three given with respect to $a + c + d$,
and the given c is resolved into $a + b + d$. But that given d is also resolved
into three given with respect to $a + b + c$. And thus only a is not resolved.
But if h were less than b, then a would be greater than m, then $a + b + c$
would be resolved into the others and $n + d$. If therefore a is not resolved,
any one will be reduced to the given in a ratio with a and the given in a
ratio with one of the others. For this is v equal to $x + s + t$. But these
equal the given ratio of $l + z + y$ to $a + b + c$. Subtract therefore y
from v and $f + e$ remains. $f + e$ will be as $x + z$ and $t + i$, $t + i$ as $n + e$,
and the ratio of $p + e$ to $a + b$ is known. Subtracting $p + e$ from $f + e$
yields $r + e$; and $r + e$ is as $t + i + p + e$. But since b is to $r + e$ and
$t + i + p + e$ is known with respect to a, then will the ratio of b to a be
known; consequently any of them will be known in respect to a. Therefore a is known with respect to $h + k + l$. And for that reason a is known
with respect to the composite z. And each of them is therefore found.

For example, let the given number be 28. And let there be four
numbers such that the first with thrice the others is 28, the second with
three and a fourth of the others is 28, the third with three and four sevenths

of the others is 28, and the fourth with four times the sum of the others is 28.[79] Subtract the first from three and a fourth, three [third and fourth] from three and a fourth, and the second from thrice itself. What remains is twice the second equal to two and a fourth of the first and a fourth of the third and fourth. For the same reason, two and a fourth of the third equals two and four sevenths of the second and nine twenty-eighths of the first and fourth. Because two and a fourth is nine times a fourth, if two and four sevenths and the nine twenty-eights are divided by nine, there obtains a fourth of the third equal to two sevenths of the second and a twenty-eighth of the first and fourth. Subtracting two sevenths [of the second] from twice [the second] leaves one and five sevenths of the second equal to two and five twenty-eighths of the first and five twenty-eighths of the fourth. Now because two and four sevenths of the fourth equals thrice the third and three sevenths of the first and second, and because two and four sevenths equals nine times eight twenty-eighths, eight twenty-eighths of the fourth equals a third of the third and a twenty-first of the first and second. And because two and a fourth of the third equals one third of six and three fourths, it follows that a third of the third equals eight twenty-firsts of the second and one twenty-first of the first and fourth. So, subtract the twenty-first of the fourth from the five twenty-eighths and what remains is twenty-four eighty-fourths of the fourth equal to nine twenty-firsts of the second and two twenty-firsts of the first. But twenty-four eighty-fourths equals eight twenty-eighths which subtracted five times yields eight twenty-eighths of the fourth, eighteen thirty-fifths of the second, and twelve one-hundred-and-fifths of the first. Since one and five sevenths of the second equals two and eight twenty-eighths of the first and eight twenty-eighths of the fourth, subtract the eighteen thirty-fifths from the one and five sevenths. There will remain one and seven thirty-fifths of the second equal to two, a fourth, a twenty-eighth, two twenty-firsts and two one-hundred-and-fifths of the first; that is, five hundred and four four-hundred-and-twentieths of the second equals one thousand and eight four-hundred-and-twentieths of the first. Since 1008 is twice 504, the second is twice the first. Now if from one and five sevenths [of the second] is subtracted two and eight twenty-eighths of the first. the remainder is eight twenty-eighths of the fourth. Since thirty-two is four times eight, it follows that the fourth is four times the first. And because two and four sevenths of the second and nine twenty-eighths of the first and fourth equal six and three fourths of the first, there will be two and a fourth of the third. Since six and three fourths equals three times two and a fourth, the third is thrice the first. Hence, thrice [the sum] of the second, third and fourth equals the ratio of 28 to one. Since they all equal 28, the first number is one, the second two, the third three, and the fourth four.

II-27.   THE ARABIC METHOD CONSISTS OF FRACTIONS AND PROCEEDS IN THIS FASHION.

For example,[80] let there be four numbers such that the first with half the others is 37, the second with a third of the others is 37, the third with a fourth of the others is 37, and the fourth with a fifth of the others is 37. Now select another set of numbers with the same fractional arrangements and set each group equal to 12 whose half is 6. Subtract from 12 the number that with a third of the remainder is 6, and it is 3.[81] Likewise subtract the next that with a fourth of the remainder is 6, and it is four. Similarly with the third that with a fifth of the remainder is 6, and it is $4\frac{1}{2}$. Add the 3, 4, and $4\frac{1}{2}$ to get $11\frac{1}{2}$ which, of course, does not equal 12. So, subtract the one from the other, which yields $\frac{1}{2}$, divide it by 3, and you have $\frac{1}{6}$. Thus for the first number is $\frac{1}{6}$, the second $3\frac{1}{6}$, the third $4\frac{1}{6}$, and for the fourth $4\frac{2}{3}$. Now the 6, which is half the three [parts] with the $\frac{1}{6}$ equals $\frac{37}{6}$. Hence, the first of the desired numbers is 1, the second is 19, the third is 25, and the fourth is 28.

  Demonstration:[82] let there be four numbers, a, b, c, and d, and let e be among the parts of the greater denomination so that with a it equals a given number. And let f equal $b + c + d$, so that a is the smallest. Let g equal $a + c + d$, which is joined with f. And since the part $c + d$ that equals e is greater than the part of those all of which is q, since b and the part a equals g, it is greater than a with part b equal to e. Subtract therefore those parts from a and b and the remainder of b will be greater than the remainder of a. Similarly, subtract the part g from a; and the remainder will be less, which means that it will be greater than the remainder of a. Hence b is greater than a, and similarly with the others. Therefore subtract from each what is equal to a, namely h, k, and l, and the remainders will be m, n, and o. Now b with its part $a + c + d$ equals $a + e$, and m with all its part equals the same e because b equals m. Therefore m with all its part $b + c + d$, that is with the whole part of its remainder in a ratio with f, equals e. Similarly with the part adjunct to c that makes

$a + e$, is there e, that is with part $k + d + b$, which is the ratio of the remainder with f and c with the part that with d equals $a + e$ will equal e, that is with part $l + b + c$ which is the ratio of the remainder to f. Subtract therefore m, n, and o from f to obtain k, h, and l. If these are divided by $b + c + d$, $h + k + l$ will equal a. Therefore we will have a, and with h, k, and l added to m, n, o, we obtain b, c, and d, which is what we wanted to do.

II–28. IN A SIMILAR MANNER THE METHOD PROCEEDS IN ANY RATIO OF SUMS, IN WHICH THE SUMS OF THE ADDENDS TO THE REMAINING PARTS ARE EITHER GREATER OR LESS THAN THE RESPECTIVE PART TO WHICH EACH IS ADDED.[83]

Assume the contrary [namely, that some sum is greater than the part to which it is added while another sum is less.][84] Given the four undefined numbers as before, w, x, y, and z, so that w plus the ratio of $e + f + g$ and $x + y + z$ equals a, x plus the ratio of $h + k + j$ and $w + y + z$ equals a, and so on. And let $e + f + g$ be greater than $x + y + z$, while $j + k + h$ is less than $w + y + z$. Call the excess of e over x, m. Consequently w with $m + f + g$ equals $h + k + j$. Because w is larger than h, so will $k + j$ be greater than $m + f + g$. But $f + g$ is greater than $y + z$, which is greater than $k + j$. This requires $f + g$ to be larger than $m + f + g$, which is impossible.

For example as before. Let the first number with thrice the sum of the others equal the given number, which shall be 28 instead of the former 12. Now let that which is added to the first number have a smaller ratio (assuming it were actually larger) to the other numbers, and let this be three. Now thrice 28 is 84. [Hence, the first number is zero.][85] Let us then subtract from 28 a number that with $3\frac{1}{4}$ of the remainder is 84; hence, this number is $3\frac{1}{9}$. Likewise, the second[86] that with $3\frac{4}{7}$ of the remainder equals 84, and this [the third number!][87] is $6\frac{2}{9}$. The third, however, with four times the remainder equals 84, and that number is $9\frac{3}{9}$. The sum of these is $18\frac{6}{9}$, which subtracted from 28 is $9\frac{3}{9}$ whose third is $3\frac{1}{9}$, the first number to be found. Adding this to the others we have $6\frac{2}{9}$ for the second, $9\frac{3}{9}$ for the third, and $12\frac{4}{9}$ for the fouth, among which the proper ratio can be found.[88]

End of Book II

Book Three

III-1. IF THE TWO EXTREMES OF THREE NUMBERS IN CONTINUED PROPORTION ARE GIVEN, THEN THE MIDDLE TERM CAN BE FOUND.
The square of the middle term is equal to the product of the extremes. Hence, take its root to find the middle term.

For example, let 9 and 4 be the extremes. Multiply them together to obtain 36 whose root is 6. This is the middle term of the continued proportion [89] between 9 and 4.

III-2. IF THE MIDDLE TERM AND ONE OF THE EXTREMES OF THREE NUMBERS IN CONTINUED PROPORTION ARE GIVEN, THEN THE OTHER EXTREME CAN BE FOUND.
If the middle term is squared and divided by the given extreme, then the other extreme is found.

For example, let 4 be an extreme and 6 the middle term. Square 6 to obtain 36, which then is divided by 4 to obtain 9. This is the third term of the continued proportion, following 4 and 6.

III-3. IF THE RATIO OF THE FIRST TO THE SECOND TERM OF THREE NUMBERS IN CONTINUED PROPORTION IS GIVEN, THEN THE RATIO OF THE FIRST TO THE THIRD CAN BE FOUND.
The ratio of the first and second multiplied by the ratio of the second and third results in the ratio of the first and third, which is the same as the square of the first (ratio).

For example, square the ratio of the first and second, $1 + \frac{1}{3}$, to get $1 + \frac{2}{3} + \frac{1}{9}$, which is the ratio of the first and third.

III-4.   IF THE RATIO OF THE FIRST AND THIRD TERMS OF THREE NUMBERS
IN CONTINUED PROPORTION IS KNOWN, THEN THE RATIO OF THE FIRST AND
SECOND CAN BE FOUND.

Since the ratio of the first to the third is the square of the ratio of the first
to the second, the root of the latter is the former.

For example, let the ratio of the first and third be $2 + \frac{1}{4}$, that is, $\frac{9}{4}$.
The root of this is $\frac{3}{2}$, which is the ratio of the first and second.[90]

III-5.   IF THE MIDDLE TERM OF THREE NUMBERS IN CONTINUED PROPORTION
AND THE SUM OF THE OTHER TWO ARE KNOWN, THEN EACH OF THESE CAN
BE FOUND.

Let the ratio of x to b equal the ratio of b to y, with b and $x + y$ known.
Since the square of b equals the product of x and y, with this product
known each of its parts can be found.[91]

For example, let the middle term be 12 and the sum of the extremes
be 26. This squared is 676. Square 12 and subtract it four times from 676.
What remains is 100, whose root is 10, the difference of the extremes.[92]
Therefore the extremes are 8 and 18.

III-6.   IF THE SUM OF THE FIRST AND THIRD TERMS IS IN A KNOWN RATIO
TO THE SECOND TERM, THE THREE BEING IN A CONTINUED PROPORTION, THEN
THE RATIO OF EACH TO THE SECOND CAN BE FOUND.

Let the ratio of $x + z$ to y be given. Now this is composed of the ratios
of x to y and z to y. Since the ratio of x to y to one equals the ratio of one
to z to y, and since the mean, one, is known, the other ratios can be
found.[93]

For example, let the ratio of the sum of the extremes with the mean
equal $2\frac{1}{12}$. The square of this is $4\frac{49}{144}$. From this subtract four times the
square of one, leaving $\frac{49}{144}$, whose root is $\frac{7}{12}$. If this is subtracted from $2\frac{1}{12}$,
there remains $1\frac{2}{4}$, whose half is $\frac{3}{4}$. Hence, the smaller term is $\frac{3}{4}$ of the
mean and the mean is $\frac{3}{4}$ of the larger.

III-7.   IF THE FIRST OF THREE NUMBERS IN CONTINUED PROPORTION AND
THE SUM OF THE OTHER TWO ARE GIVEN, THEN BOTH CAN BE FOUND.

Let the continually proportional numbers be a, y, and z, with a and $y + z$
given. Let the product of a and $y + z$ be $d + e$ so that ay is d and az is e.

158

The English Translation

e is also *y* squared. So the sum of *y* squared and *y* times *a* is known. Therefore the parts can be found.[94]

For example, let one of the extremes be 9 and the sum of the others be 28. Multiply 9 by 28 to get 252, which quadrupled is 1008. To this add the square of 9 to obtain 1089. The root of this is 33, from which 9 is subtracted to yield 24. Half of this is 12, the middle term. The third term is 16.

III-8.   IF THE RATIO OF THE SUM OF AN EXTREME AND THE MEAN TO THE OTHER EXTREME IS GIVEN, THEN THE RATIO OF EACH TO THE MEAN CAN BE FOUND.

Let the ratio of $x + y$ to z be given, and this is the sum of the ratios of x to z and of y to z. Now the ratio of x to z is to the ratio of y to z as the ratio of y to z is to one. By the previous method,[95] the desired ratios can be found.

For example, let the given ratio be 6, which taken four times is 24. Add one to this to obtain 25, whose root is 5. Subtract one from this and halve the remainder to get 2. Hence, the mean is twice the lesser term and half the greater.

III-9.   IF THE THIRD TERM OF A CONTINUED PROPORTION IS GIVEN TOGETHER WITH THE SUM OF THE FIRST AND TWICE THE SECOND, THEN THESE TWO CAN BE FOUND.

Let x with twice y be the given number, and let b be known. Then square b to get d, multiply b by x to get e, b by twice y to get $f + g$. Therefore $d + e + f + g$ is known, because e is equal to the square of y. Hence, $d + e + f + g$ is equal to the square of $y + b$. So take its root to find $y + b$. Since b is known, y is found as well as x.[96]

For example, let one extreme be 2 and the sum be 16. Therefore the square of 2 and the product of 2 and 16 added together are 36. Take the root to get 6, from which 2 is subtracted to obtain 4, the mean. The third term is 8.

III-10.   LET THREE NUMBERS BE CONTINUALLY PROPORTIONAL. IF THEIR SUM IS KNOWN AND THE RATIO OF THE EXTREMES IS GIVEN, THEN EACH OF THEM CAN BE FOUND.

For if the ratio of the extremes is given, then the ratio of the first to the second and the second to the third can be found.[97] If the sum is then divided proportionally, the three terms can be found.

For instance, let the sum of the three be 19 and the ratio of the extremes be $2\frac{1}{4}$. The root of $2\frac{1}{4}$ is $1\frac{1}{2}$. Divide 19 by the three proportionally so that the first is $\frac{2}{3}$ of the second, which is $\frac{2}{3}$ of the third, namely 4, 6, and 9.

III-11.   IF THE SUM OF THREE NUMBERS IN A CONTINUED PROPORTION ARE GIVEN TOGETHER WITH THE DIFFERENCE OF THE EXTREMES, THEN ALL THREE TERMS CAN BE FOUND.

Let the given difference be added to and subtracted from the sum, so that the sum and the remainder are known, namely twice the largest with the mean, and twice the smallest with the mean. Multiply these two together. Now since twice the smallest times twice the largest is four times the mean, it follows that three times the square of the mean plus twice the product of the mean and the sum of the three is known. Hence, a third of all this is known. Consequently the mean can be found as well as the other two terms.[98]

For example, let the sum of the three be 38 and the difference of the extremes be 10. First subtracting, then adding it to 38 yields 28 and 48. Multiplying these two together gives 1344, whose third is 448. Quadruple this to obtain 1792. To this add the square of two-thirds of 38, namely $25\frac{1}{3}$, to make $2433\frac{2}{3} + \frac{1}{9}$, whose root is $49\frac{1}{3}$. From this subtract $25\frac{1}{3}$. Halve the remainder to obtain 12, which is the mean. The other two parts equal 26, of which one is 8 and the other is 18.

III-12.   IF THE SUM OF THE TWO EXTREMES OF THREE NUMBERS CONTINUALLY PROPORTIONAL AND THE SUM OF THE MEAN AND THE SMALLER EXTREME ARE GIVEN, THEN THE THREE NUMBERS CAN BE FOUND.

Let x be the largest of three numbers, x, y, and z, in a continued proportion. Let the sums of x and z and of y and z be given. Call half the difference of x and z, d. [99] Now it is obvious that $z + d$ is half of $x + z$. Moreover, the square of $z + d$ is equal to the sum of the squares of y and d, because the square of $z + d$ equals the square of d plus the products of d and z and of $z + d$ and z. The sum of $z + d$ and d is x, and the product of x and z is y squared. Now since $y + z$ is known as is $z + d$, their difference can be found, which is the difference of d and y. Hence, since the squares of those are known, their sum as well as each of them can be found. Since y has been found and $x + z$ is known, x and z are found.

For example, let the sum of the extremes be 34 and the sum of the mean and lesser extreme be 24. Now the square of half 34 is 289. This is equal to the square of the mean plus the square of half the difference of the extremes, whose difference is 7. Square the 7 to get 49, which in turn is subtracted from 289 to yield 240. Add this to the other to get 529, whose root is 23. From this subtract 7. Halve the remainder to obtain 8, which is subtracted from 23 to get 15, the mean. Thus the two extremes are 9 and 25.

III-13.   IF THE SUMS OF THE EXTREMES AND OF THE MEAN AND THE LARGEST OF THE THREE TERMS ARE GIVEN, THEN THERE ARE TWO WAYS TO DETERMINE THE TERMS.

Given $x + z$ and $x + y$, there are two ways to find x, y, and z. Let d be half the difference of x and z, with e the greater part and f the smaller. $x + z$ will always be greater than the square[100] of y. Now I say that d will be the mean proportional between z and f and that $z + d$ equals $f + y$. Since $f + y$ is half of $x + z$, it follows that its square equals the sum of the squares of y and d; therefore the square of d equals the sum of the products of y and f and of y and the square of f. And because $f + y$ and likewise y are as e, the square of d equals the product of e and f. Hence d is (the mean proportional) between e and f. Since $e + d$ equals half of y and d, and $x + y$ similarly equal, they are therefore known, as are d and y. Since $x + y$ was given and its half is known, so is $d + y$ found. But their squares are known, hence each is found. Therefore both e and f together with x and z are found.

For example, let the sum of the largest term and the mean be 28 and the sum of the extremes be 25. Half of 25 is $12\frac{1}{2}$ and its square is $156\frac{1}{4}$. Subtracting the half from 28 leaves $15\frac{1}{2}$, whose square is $240\frac{1}{4}$. Subtract $156\frac{1}{4}$ from this to get 84, whose difference with $156\frac{1}{4}$ is $72\frac{1}{4}$. The root of this is $8\frac{1}{2}$, which subtracted from $15\frac{1}{2}$ is 7, which is halved in turn to yield $3\frac{1}{2}$. This with 12 makes the mean: for either 12 or $3\frac{1}{2}$ can be the mean. If it is 12, then the extremes are 16 and 8. If it is $3\frac{1}{2}$, then they are $24\frac{1}{2}$ and $\frac{1}{2}$.

III-14. IF FOUR NUMBERS ARE IN PROPORTION SUCH THAT THE FIRST AND FOURTH AND THE SUM OF THE SECOND AND THIRD ARE KNOWN, THEN EACH CAN BE FOUND.

Since the first and fourth are known, their product equals the product of the second and third. Since the sum of the second and third is also known, each can be found.[101]

For example, let the first term be 15, the fourth 6, and the sum of the second and third 19. The product of 15 and 6 is 90 and the square of 19 is 361. From this take four times 90 to get one. The root of this is one, which is the difference of the second and third. Although they are 10 and 9, it is not clear which is which.[102]

III-15. IF FOUR NUMBERS ARE IN PROPORTION SUCH THAT THE FIRST AND FOURTH AND THE DIFFERENCE OF THE SECOND AND THIRD ARE KNOWN, THEN EACH CAN BE FOUND.

For the same reason as above, the product of the second and third is known which, together with the given difference, produces each term.[103]

For example, let the first be 12, the fourth 3, and the difference of the

to the square of 5 makes 169. The root of this is 13, from which 5 is subtracted. Halve the remainder to get 4, which is one of the terms, and the other is 9. Again it is not definite which is which.

III-16. LIKEWISE, IF THE FIRST AND FOURTH TERMS ARE GIVEN AND ALSO THE RATIO OF THE SECOND TO THE THIRD, THEN EACH OF THESE CAN BE FOUND.

Since the first and fourth are given, their ratio is known. This consists of the ratio of the first to the third times the third to the second times the second to the fourth.[104] Now divide the given ratio of the second and third by the ratio of the fourth and first, and this quotient is known; moreover it is made up of the ratio of the first to the third and of the second to the fourth.[105] Take the root of this to get the ratio of the first to third; and therefore the third is found. Similarly the ratio of the second to the fourth is determined, and the second term is found.

For example, let the first term be 18, the fourth 2, and the ratio of the second and third equal to 4. Now 18 is 9 times 2, so divide 9 by $\frac{1}{4}$ to obtain 36. Take the root of this to get 6. Therefore the ratio of the first and third is 6 and the third term is 3. And the ratio of the second and fourth is 6 which makes the second term 12.

III-17. IF THE FIRST AND FOURTH OF FOUR NUMBERS IN PROPORTION AND THE RATIO OF THE SUM OF THE FIRST AND SECOND TO THE THIRD ARE GIVEN, THEN EACH CAN BE FOUND.

Let the proportional numbers be a, x, y, and d, and let a, d, and the ratio of $a + x$ to y be given. Now the ratio of $a + x$ to y consists of the ratio of $a + x$ to a (times) the ratio of a to y. The ratio of $a + x$ to a is the same as x to a plus one. Consequently the ratio of a to y times the ratio of x to a plus one equals the ratio of $a + x$ to y. But the ratio of a to y times the ratio of y to d equals the ratio of a to d. Therefore the ratio of a:d to $a + x$:y equals the ratio of y:d to x:$a + 1$.[106] Since the ratio of y:d to one equals the ratio of one to x:a, namely the ratio of each to the mean is known, it follows that each can be found. And these yield x and y.[107]

For example, let the first term be 16, the fourth 3, and the ratio of the first and second to the third be 4. Now the ratio of 16 to 3 is $5\frac{1}{3}$ and the ratio of $5\frac{1}{3}$ to 4 is $1 + \frac{1}{4} + \frac{1}{12}$. 4 is $\frac{3}{4}$ of $5\frac{1}{3}$. To $\frac{3}{4}$ of 4 add one to get 4 whose root is 2. From this subtract 1, which is halved to obtain $\frac{1}{2}$. Therefore the second term is half of 16 or 8, and the third is twice 3 or 6.

Another way to do it is this. Take a fourth of 16 or 4 as $\frac{1}{3}$ of the first and second, and multiply 3 by 4 to get 12. Quadruple this and add it to the square of 4 to obtain 64. The root of this is 8, from which 4 is subtracted. Halve the remainder to yield 2 that with 4 makes 6, the third term. The second term is 8.

III-18. IF THE SUM OF THE FIRST TWO OF FOUR NUMBERS IN PROPORTION, THE SUM OF THE SECOND TWO, AND THE RATIO OF THE FIRST AND FOURTH ARE GIVEN, THEN EACH NUMBER CAN BE FOUND.

Since the sums are given and they form a known ratio, then the ratio of the first to the third is also known. Since the ratio of the first and fourth is known, then the ratio of the first to the sum of the third and fourth is known. Consequently the first can be found, as well as the other three.

For example, let the sum of the first two be 25, of the second two be 10, and the ratio of the fourth and first be $\frac{4}{15}$. Since 10 is $\frac{2}{5}$ of 25, then the ratio of the sum of the third and fourth to the first is $\frac{10}{25}$. Since the sum of the third and fourth is 10, the first term is therefore 15, the second is 10, the third is 6, and the fourth is 4.

III-19. IF THE SUMS OF THE EXTREMES AND THE MEANS TOGETHER WITH THE RATIO OF THE FIRST AND THIRD TERMS ARE KNOWN, THEN EACH TERM CAN BE FOUND.

Let there be given the sums $w + z$ and $x + y$ and the ratio of w and y. Consequently the following can be found: the ratio of $w + x$ to $y + z$, the sum of $w + x + y + z$, and the sums of $w + x$ and $y + z$. Also are known the differences of x and z and of w and y. Now, since the ratio of the differences of w and y to x and z is the same as the ratio of $w + y$ and $x + z$, and knowing the sum of $w + x + y + z$, $w + y$ and $x + z$ can be found. Since the differences of w and y and of x and z have been found, all the terms can be found.

For example, let the sum of the extremes be 16, the sum of the means be 14, and the ratio of the first and third terms be $\frac{3}{2}$. After a combination of one and one and a half,[108] that is, 30 the sum of all (the terms), (the ratio of this) to the sum of the third and fourth will be $2\frac{1}{2}$. Therefore it (the latter sum) will be 12. But the fourth with the first was 16; therefore the first exceeds the third by 4. Hence the fourth is half the third, which is therefore 8. And the first term is 12, the second 6, and the fourth 4.

III-20. IF OF FOUR NUMBERS IN PROPORTION, THEIR SUM IS GIVEN TOGETHER WITH THE DIFFERENCES OF THE FIRST AND SECOND AND OF THE THIRD AND FOURTH TERMS, THEN ALL THE NUMBERS CAN BE FOUND.

For if the differences of the first and second and of the third and fourth are given, then the difference of the sums of the first and third and of the second and fourth are known. These sums are known, since the sum of all four is known. But the ratio of the two sums equals the ratios of the first to the second and of the third to the fourth, which makes these ratios known. Therefore the ratio of the first to the difference of the first and second, and the ratio of the third to the difference of the third and fourth are known. Hence, these numbers and the others can be found.

For example, let the sum of the numbers be 35, the difference of the first and second be 5, and of the third and fourth be 2. Therefore the difference of the first and third with the second and fourth is 7, which is subtracted from 35 to obtain a remainder that is halved to yield 14, which equals the second and fourth. Therefore the sum of the first and third is 21. Since this is thrice 7, the difference between 21 and 14, it follows that the ratio of the first is 5 (the difference of the first and second) and of the third to 2 (the difference of the third and fourth) is 3. Hence, the first term is 15, the second is 10, the third is 6, and the fourth is 4.

III-21.   IF THE SUM OF FOUR NUMBERS IN PROPORTION IS GIVEN TOGETHER WITH THE DIFFERENCES OF THE FIRST AND FOURTH AND OF THE SECOND AND THIRD, THEN ALL THE NUMBERS CAN BE FOUND.

Let the sum of w, x, y, and z be given. Let e be the given difference of w and z, and h the given difference of x and y. Let w be the largest number and x greater than y. Since the differences are given, if e is subtracted from h, what remains is the difference of w and x added to the difference of y and z; and this is known. Since two differences with the sum total were given, then $w + y$ and $x + z$ can be found. Moreover, since the difference of w and y consists of the sum of h and the difference of w and x, similarly the difference of x and z consists of the sum of h and the difference of y and z (and these four differences are as e and h), then e and h produce the difference of $w + x$ and $y + z$, which therefore can be found. Now the ratio of $w + x$ and $y + z$ equals the ratio of w to y and of x to z, which therefore can be found. And since $w + y$ and $x + z$ are known, w and y can be found as well as x and z.

For example, let the total sum be 45, the difference of the first and fourth be 7, and the difference of the second and third terms be 2. The difference of the differences is 5, which then is subtracted from 45. Half the remainder is 20, the sum of the second and fourth. The sum of the first and third is 25. Now add the 2 to 7 to get 9, which is subtracted from 45. Halve the remainder to yield 18, the sum of the third and fourth terms. 27 is the sum of the first two terms. Now since 25 is greater than 20 by its fourth, the first term is five-fourths of the second. Likewise, since 27 is $\frac{9}{4}$ of the second, the latter equals 12. Hence the first term is 15, the third is 10, and the fourth is 8.

III-22.   IF THREE NUMBERS IN CONTINUED PROPORTION ARE COMPARED WITH THREE OTHER NUMBERS IN CONTINUED PROPORTION SO THAT THE RATIOS OF THE FIRST AND OF THE THIRD TERMS ARE GIVEN, THEN THE RATIO OF THE MEANS CAN BE FOUND.

Let the first set be u, v, and w, the second set be x, y, and z, so that the ratios of u to x and of w to z are given. Then the ratio of v to y can be found. Find the product of the ratios of u to x and of w to z, and take its root. This is the ratio of v to y.

For example, let the ratio of the first terms be $\frac{9}{8}$, of the third terms be 2, and their product be $\frac{18}{8}$, which is the denomination of the composite ratio, were it continued. The root of this is $\frac{12}{8}$, the ratio of the means, for the ratio of the product of the extremes equals the square of the ratio of the means.

III-23. IF HOWEVER SO MANY NUMBERS IN CONTINUED PROPORTION ARE COMPARED WITH ANOTHER LIKE SET OF NUMBERS AND THE RATIOS OF THE FIRST AND OF THE SECOND TERMS ARE GIVEN, THEN THE RATIOS OF THE REMAINING PAIRS CAN BE FOUND.

For the quotient[109] of the ratio of the respective first terms to the ratio of the respective second terms equals the quotient of the ratios of the first and second term of one set and of the first and second terms of the second set. And this in turn equals the quotient of the ratios of the second and third terms of the first set and of the second and third terms of the second set, and so on. Now this last quotient equals the quotient of the ratios of the respective second and third terms. Hence, in continued proportion the quotient of the ratios of the respective first terms to the second terms equals the quotient of the ratios of the respective second terms to the third terms. One may then proceed in a similar fashion dividing and multiplying until all the terms have been exhausted. Therefore by continually dividing the quotients, the ratio of the remaining terms to one another can be found.

For example, compare two sets of four terms. The ratio of the respective first-terms is $\frac{4}{3}$ and of the second terms is 1. Therefore divide $\frac{4}{3}$ by 1, which has denominated equality. The quotient is $\frac{4}{3}$. Now divide 1 by $\frac{4}{3}$ and the quotient is $\frac{3}{4}$. Therefore the ratio of the [respective] third terms is $\frac{3}{4}$. To find the ratio of the fourth terms, divide $\frac{3}{4}$ by $\frac{4}{3}$ and the answer is $\frac{9}{16}$.

End of Book III

Book Four

IV-1. IF TWO NUMBERS ARE DIVIDED BY TWO OTHERS AND THE RATIOS OF THE DIVISORS AND OF THE DIVIDENDS ARE KNOWN, THEN THE RATIO OF THE [ORIGINAL] QUOTIENTS CAN BE FOUND.

Let w and x be divided by y and z to produce m and v, and let the ratios of w to x and of y to z be given. Now divide the given ratios by one another and this produces the ratio of m and v, for the ratio of w and x equals the product of the ratios of y to z and of m to v.

For example, let the ratio of the divisors be 2 and the ratio of the dividends be 3. Dividing 3 by 2 yields $1\frac{1}{2}$. Hence the ratio of the quotients is $1\frac{1}{2}$.[110]

IV-2. IF THE RATIOS OF THE DIVISORS AND OF THE QUOTIENTS ARE GIVEN, THE RATIO OF THE DIVIDENDS CAN BE FOUND.

For a similar reason [as above] by multiplying one ratio by the other, the desired ratio is found.

For example, let the ratio of the quotients by $1\frac{1}{2}$ and the ratio of the divisors be $1\frac{1}{3}$. The product of these two numbers is 2; hence, the ratio of the dividends is 2.

IV-3. IF A GIVEN NUMBER IS DIVIDED BY TWO NUMBERS WHOSE DIFFERENCE IS KNOWN AND THE DIFFERENCE OF THE QUOTIENTS IS ALSO GIVEN, THEN THE NUMBERS CAN BE FOUND.

Let the given number be a, which is to be divided by x and y,[111] whose given difference is d. Let the respective quotients by z and w, whose known difference is g. And thus let x be to d as h is to y. But since x is to *d* as *w* is to *g*, it follows that *w* is to *g* as *h* is to *y*. Now *w* times *y* equals *a*; therefore *h* times *g* is *a*. Likewise *x* times *y* is *L*; therefore *d* times *h* is *L*. And so *a* is to *L* as *g* is to *d*; hence *a* times *d* divided by *g* produces *L* as known. Wherefore since the difference of *x* and *y* was given, so can *x* and *y* be found[112] and thence *z* and *w*.

For example, divide 24 by two numbers whose difference is one and let the difference of the respective quotients be 2. Now one times 24 is 24, and this divided by 2 is 12. By quadrupling this and adding it to the square of one, 49 is obtained, whose root is 7. From this subtract one; then halve the remainder to obtain 3, which is the smaller of the divisors; the larger is 4. Hence, the quotients are 8 and 6.

IV-4. IF THE DIFFERENCE OF THE QUOTIENTS AND THE SUM OF THE DIVISORS ARE GIVEN, THEN EACH OF THEM CAN BE FOUND.
Let there be given the information from the above theorem, except that *x* + *y* is known and not *d*; moreover let the ratio of *a* to *g* be as *L* to *d*. Now if *L* is divided by *d*, a certain thing is known. And since *d* is the difference of *y* and *x* (which makes one known) and *L* is the product of *x* and *y*, then both *y* and *x* are found, as are *z* and *w*.[113]

For example, let 20 be divided by two numbers whose sum is 7 and let the difference of the quotients be 6. Now, divide 20 by 6 to obtain $3\frac{1}{3}$. Quadruple this to obtain $13\frac{1}{3}$. Square this and add it to four times the square of 7. The sum is $373\frac{7}{9}$, whose root is $19\frac{1}{3}$. From this subtract $13\frac{1}{3}$ and halve the remainder to obtain 3. Subtract this from 7 and halve the remainder to obtain 2, which is one of the divisors, the other being 5. The quotients are 10 and 4.

IV-5. IF THE RATIO AND PRODUCT OF THE SAME TWO NUMBERS ARE GIVEN, THEN EACH OF THEM CAN BE FOUND.
Let the ratio of *x* and *y* be known and call their product *b*. Therefore let some number be to *b* as *x* is to *y*. Call this *z* and it is found; moreover it is equal to the square of *x*. Take its root to find *x*. Then *y* can be found.

For example, let the ratio of the two numbers be $1\frac{1}{3}$ and their product be 48. Add to 48 its third to obtain 64, whose root is 8. This is one of the numbers; the other is 6.

IV-6. IF THE RATIO OF TWO NUMBERS TOGETHER WITH THE SUM OF THEIR SQUARES IS KNOWN, THEN EACH CAN BE FOUND.

Let the ratio of x and y be given. Call b the square of x and c the square of y; and $b + c$ is known. Now the ratio of b to c is the square of the ratio of x and y; hence the former is known. Consequently b and c can be found.

For example, let the ratio of two numbers be 2 and the sum of their squares 500. Now since the square of one is four times the square of the other, it follows that 500 is five times the square of the other, which makes it 100. The root of this is 10 for the smaller number and the larger is 20.

IV-7.   IF THE RATIO OF TWO NUMBERS IS GIVEN TOGETHER WITH THE PRODUCT OF THEIR SUM AND DIFFERENCE, THEN EACH CAN BE FOUND.[114] Now the product of the sum and difference equals the difference of their squares.[115] Moreover, since the ratio of the squares of the two numbers is known because the ratio of one number to the other is given, one of the squares can be found. Therefore its side[116] is known and the other can be found.

For example, let the ratio be 3 and the product of their sum and difference be 32. Now the square of the ratio is 9. So the sum is 8 times one of the squares. Wherefore the square of the less is 4, making it 2 and the other 6.

IV-8.   IF THE SUM IS KNOWN OF THE SQUARE OF A NUMBER AND THE PRODUCT OF A GIVEN NUMBER AND THE ROOT OF THE SQUARE, THEN THE NUMBER CAN BE FOUND.[117] Let the square be a, multiply its root b by $c + d$ (c and d each being half of $c + d$) to obtain e, and let $a + e$ be given. Now since $b + c + d$ multiplied by b equals $a + e$, by adding the square of d to $a + e$ there obtains $a + e + f$. Now $a + e + f$ equals the square of $b + c$.[118] Since $a + e + f$ was given, then $b + c$ can be found. By subtracting c, b is found and then a is found.[119]

For example, let 36 be the sum of the square and 5 times the root. To this add the square of $2\frac{1}{2}$ (which is half of 5) to obtain $42\frac{1}{4}$. The root of this is $6\frac{1}{2}$. From this subtract $2\frac{1}{2}$ to obtain 4, which is the root. The square is 16.

IV-9.   IF THE SUM OF THE SQUARES OF A NUMBER AND A GIVEN NUMBER EQUALS THE PRODUCT OF THE ROOT AND ANOTHER GIVEN NUMBER THEN IT HAS TWO POSSIBLE VALUES.[120]

Let the unknown number be x and the given addend be c and let $d + c$ be the factor which multiplied by x equals $x^2 + c$. Moreover, let d be half of $d + c$, its square be f, and the difference of x and d be g. Now since x times twice d equals $x^2 + c$, x^2 and f equal $x^2 + c$ plus the square of g. Subtracting x^2 from both sides of the equation leaves f equal to c plus the square of g. Therefore by subtracting g from d, it is possible to obtain x; and be adding g to d, it is possible to obtain x. Hence there are two possible values for x.

For example, let the square plus 8 equal 6 times the root. Square half of 6 to get 9, which exceeds 8 by one. The root of this is one and it is the difference of the root and 3. By subtracting and adding this difference to 3 we have 2 and 4, whose squares are 4 and 16. To each add 8 to obtain 12 and 24, which can also be found by multiplying each by 6, as was proposed.

IV-10.   IF THE PRODUCT OF A GIVEN NUMBER AND AN UNKNOWN IS ADDED TO ANOTHER GIVEN NUMBER AND THEIR SUM EQUALS THE SQUARE OF THE UNKNOWN, THEN IT CAN BE FOUND.[121]

As before let x be the unknown, $c + d$ the factor, and e the addend. Now the difference of x and $c + d$ is g, so that $g + c + d$ equals x. And because the square of x is x^2, which also equals e plus the product of x and $c + d$, it is obvious that e equals x times g. Now since the square of $g + c$ is as much as the product of x and g and the square of d, this also equals the square of d increased by e. Since this was given, so will $g + c$ be found. Hence g is known as is $g + c + d$, which is equal to x.

For example, the square of the unknown equals 4 times the root plus 12. Take half of 4 and square it to get 4, increase this by 12 to obtain 16, take its root which is 4, subtract half of it[122] from itself to yield 2, which being added to 4 produces 6, the root.

IV-11.   IF A NUMBER IN A GIVEN RATIO TO A SQUARE IS ADDED TO ANOTHER NUMBER WITH A GIVEN RATIO TO THE ROOT OF THE SQUARE TO MAKE A GIVEN NUMBER, THEN THE SQUARE AND THE ROOT CAN BE FOUND.

Let x be the root and y the square, and let u be in a given ratio to x and v to y, and let $u + v$ be given. Then let y be to v as g is to u. Thus $g + y$ is to $u + v$ as y is to v; hence, $y + g$ is found. Now g to x is found and call this ratio e. Wherefore x times e equals g, which with the square y equals the given number.[123] Therefore x and y can be found.

For example, let $\frac{1}{3}$ of a root and $\frac{1}{4}$ of a square equal 11. Therefore the square with $\frac{4}{3}$ of the root equals 44. To this add the square of $\frac{2}{3}$, which is half of $\frac{4}{3}$, and it becomes $44\frac{4}{9}$, whose root is $\frac{20}{3}$, or $6\frac{2}{3}$. From this subtract $\frac{2}{3}$ to obtain 6, which is the unknown whose square is 36.

IV-12. IF THE SUM OF A KNOWN NUMBER IN A RATIO TO A SQUARE AND OF A GIVEN NUMBER EQUALS THE RATIO OF ANOTHER KNOWN NUMBER IN A RATIO WITH THE ROOT OF THE SQUARE, THEN THE ROOT AND SQUARE CAN BE FOUND.

As before let x be the root and y the square and let v be in a given ratio to y that with a known c makes $c + y$ known in a ratio to x. Since therefore v is to y, so is e to c; whence e is found, and $y + c$ is a known number in a ratio to x. Therefore and similarly x and y can be found.

For example, let a twentieth of the square with 25 equal three roots. Thus the square with 500 equals 60 roots. The square of half of 60 is 900, which is 400 more than 500. The root of 400 is 20, which is the difference between the root of the square and 30. By adding to, then subtracting, this from 30 you obtain two roots, 10 and 50, whose squares are 100 and 2500. Taking a twentieth of each of these, namely 5 and 125, and adding 25 to each, you obtain on the one hand 30 (which is thrice 10), and on the other 150 (which is thrice 50), as was desired.

IV-13. IF THE SUM OF A KNOWN NUMBER IN A RATIO TO A ROOT AND OF ANOTHER KNOWN NUMBER EQUALS THE RATIO OF A GIVEN NUMBER AND THE SQUARE, THEN THE ROOT AND THE SQUARE CAN BE FOUND.

Let the arrangement be as above except that v is known in a ratio to x and the sum $c + v$ is given in a ratio to y. In the same manner as y is to $c + v$ so is e to c and f to v; hence e is found. Now f is known in a ratio to x, and $e + f$ equals y. Therefore from this can x and y be found.

For example, let thrice the root with 12 equal three halves of the square. Therefore, twice the root and 8 equal the square. After the manner of proposition 10 above, the unknown is found to be 4.[124]

IV-14. IF THE RATIO OF THE SUM OF TWO NUMBERS TO A THIRD IS KNOWN AS IS THE RATIO OF THE PRODUCT OF THE TWO TO THE SQUARE OF THE THIRD, THE RATIOS OF EACH OF THE TWO TO THE THIRD CAN BE FOUND.

Let the ratio of $x + y$ to z be given, and let the square of z be v and the product of x and y be u, so that the ratio of u and v is known. Let the ratio of $x + y$ to z be $f + g$, and this ratio is the sum of the ratios of x to z and of y to z. Let the ratio of u to v be b, and this is composed of the products of f and g. Now since $f + g$ is known and f times g is known, f and g can be found.[125] Therefore the ratios of x to z and of y to z can be found.

For example, let the ratio of the sum of two numbers to a third be 5 and the ratio of their product to the square of the third be 6. Now subtract 6 four times from the square of 5 and the remainder is one whose root is one. Subtract this from 5 and halve the remainder to obtain 2. The first root is then thrice the third, and the second is twice the third.[126]

IV-15.   IF THE RATIO OF THE SUM OF TWO NUMBERS TO A THIRD IS GIVEN TOGETHER WITH THE RATIO OF THE SUM OF THE SQUARES OF THE SAME TWO TO THE SQUARE OF THE THIRD, THE RATIO OF EACH OF THE TWO TO THE THIRD CAN BE FOUND.

Let the previous information[127] be kept except that the squares of x, y f, and g are e, t, h, and l, respectively. The ratio of e to d will be h and of t to d, l; hence, $h + l$ will be found. And since $f + g$ is given, then both f and g are found. Consequently, the ratios of x and y to z will be found.

For example, let the simple ratio be 3 and the ratio of the squares be $4\frac{5}{9}$. Double the latter to get $9\frac{1}{9}$, from which the square of 3 is subtracted to yield $\frac{1}{9}$. The root of this is $\frac{1}{3}$. Subtract this from 3 and halve the remainder to obtain $\frac{4}{3}$. Therefore one of the two is $\frac{4}{3}$ of the third and the other is $\frac{2}{3}$.

IV-16.   LIKEWISE, GIVEN THE RATIO OF THEIR[128] DIFFERENCE TO A THIRD NUMBER TOGETHER WITH THE RATIO OF THEIR PRODUCT TO THE SQUARE OF THE THIRD, THEN THE RATIO OF EACH TO THE THIRD CAN BE FOUND.

Let the difference of x and y be q; the rest can remain as in the previous theorems. Likewise, let the difference of f and g be p, and that is the ratio of q to z. Now let the product of f and g be h, which is known. Their difference is also known. Therefore they are known. Hence, the ratios of x and y to z are found.

For example, let the difference of the two equal the third, and the product of the two equal six times the square of the third. To four times the second ratio add the square of the first. This gives 25, whose root is 5.

From this subtract 1, and halve the remainder to obtain 2. Hence, the ratio of one number to the third is 2, of the other to the third is 3.

IV-17. IF THE RATIO OF THEIR DIFFERENCE TO THE THIRD IS KNOWN TOGETHER WITH THE RATIO OF THE SUM OF THEIR SQUARES TO THE SQUARE OF THE THIRD, THEN THE RATIO OF EACH TO THE THIRD CAN BE FOUND.
As before, let the difference of x and y be q and the difference of f and g be p, which is the ratio of q to z and y as in the previous theorems. Now since $h + l$, which are the square of f and g, is given, and their difference is given, then they are found.

For example, let the ratio of the difference to the third be 2 and the ratio of the squares be 20. Double the 20 to get 40, from which the square of 2 is to be subtracted. The remainder is 36 and its root is 6. Subtract 2 from this and halve the remainder to get 2. Hence 2 is the ratio of one pair of numbers, 4 the ratio of the other.

IV-18. IF THE RATIOS OF TWO NUMBERS TO A THIRD ARE GIVEN TOGETHER WITH THE RATIO OF THEIR PRODUCT TO THE SAME THIRD, THEN EACH OF THEM CAN BE FOUND.
Let the ratios of x and y to z be given. Call the product of x and y, d; and let the ratio of d to z be given. Let the square of z be e. Now if the ratio of x to z is multiplied by the ratio of y to z, there results the ratio of d to e. If this divides the ratio of d to z, the ratio of e to z, which was denominated by it, results. Hence, z is found.

For example, let the first ratio be $1\frac{1}{2}$, the second be $1\frac{1}{3}$, and the ratio of the product and the third number be 84. Multiply $\frac{3}{2}$ by $\frac{4}{3}$ to get two, which divides 84 to produce 42, the number that was sought.[129]

IV-19. IF THE PRODUCT OF TWO NUMBERS IS GIVEN TOGETHER WITH THE SUM OF THEIR SQUARES, THEN EACH OF THEM CAN BE FOUND.
Let x and y be the numbers,[130] c and d their squares. Let e be the given product of x and y, and $c + d$ is given. Now since e is the mean proportional between c and d, and since it and the sum are given, then the numbers can be found.

For example, let the product be 35 and the sum of the squares be 74. Square the latter to obtain 5476. From this subtract the square of 35,

four times, to obtain 576. The root of this is 24, which subtracted from 74 leaves 50. Halve this to get 25, which is the smaller of the squares, the other being 49. Hence, the numbers are 5 and 7.

IV-20.   IF THE PRODUCT OF TWO NUMBERS IS GIVEN TOGETHER WITH THE DIFFERENCE OF THEIR SQUARES, THEN EACH OF THEM CAN BE FOUND.
Utilize the previous information. Let g be the given difference of c and d. Since e is the mean proportional of c and d, c is to e as e is to d. Hence, c is the first term and d is the fourth of the given. And e, the second and third term, is their given difference. Hence, all of them can be found. Or, to prove it in another way: since e is known, the product of c and d is e squared. Since their difference is known, so are they.

For example, let the product be 15 and the difference of the squares 16. Square the product to obtain 225, multiply this by 4, and add it to the square of 16 to obtain 1156. The root of this is 34. From this subtract 16, halve the remainder, and the result is 9, which is the square of the smaller number. The larger square is 25. The two numbers are 3 and 5.

IV-21.   IF THE SUM OF TWO NUMBERS IS GIVEN TOGETHER WITH THE PRODUCT OF THEIR SQUARES, THEN EACH OF THEM CAN BE FOUND.
Take the root of the product and that will be the product of the sides. Since that is found and the sum of the two numbers is given, the two can be found.

For example, let the sum of the numbers be 9 and the product of their squares be 324. The root of this is 18, which is the product of the numbers. Since their sum is given, the numbers can be found as before to be 3 and 6. For by subtracting 18 four times from the square of 9, the remainder becomes 9, which is the square of the difference of the same numbers.[131]

IV-22.   LIKEWISE, IF THE DIFFERENCE OF THE NUMBERS AND THEIR PRODUCT ARE GIVEN AS BEFORE, THEN EACH OF THEM CAN BE FOUND.
In a similar fashion take the root of the product of the squares to obtain the product of the numbers. Since the difference has been given, the numbers can be found.

For example, let the product of the squares be 100; its root is 10. Let the difference of the two numbers be 3. Square this, add it to 10 four times to get 49. Take the root, which is 7, and the sum of the sides. These are, therefore, 2 and 5.

IV-23.   LIKEWISE, IF THE DIFFERENCE OF THE NUMBERS AND THE DIFFER-
ENCE OF THEIR SQUARES ARE KNOWN, THEN EACH OF THEM CAN BE FOUND.[132]
Divide the differences of the squares by the difference of the roots and one obtains the sum of the roots.[133] Since this is now known and the difference of the roots was given, the numbers can be found.

For example, let the difference of the roots be 3 and the difference of the squares be 51. Divide the one by the other and the result is 17, the sum of the roots. Subtract 3 from that and halve the remainder to obtain 7, the smaller number. The larger is 10.

IV-24.   IF THE RATIOS OF THE SQUARES OF TWO NUMBERS TO ONE ANOTHER ARE GIVEN, THEN EACH OF THEM CAN BE FOUND.
Let the two numbers be x and y. Let the ratios of x^2 to y and of y^2 to x be given. Call the first g and the second e. Now e times x equals y^2. There-fore e is to y as y is to x. Similarly, since g times y equals x^2, g is to x as x is to y. Thus the four terms are in continual proportion; namely, e, y, x, and g. Of these e and g are known. If therefore the cube root is taken of the ratio of e to g, one obtains the ratio of e to y. Hence, y is found. If, therefore, g is multiplied by the square of e and the cube root of that product is found, it is y. x can be found in a similar fashion.

For example, let one ratio be $5\frac{1}{3}$ and the other be 18. Now multiply 18 by the square of $5\frac{1}{3}$ (namely, $28\frac{4}{9}$). The result is 712. The cube root of this is 8, one of the two numbers. The other is 12.

IV-25.   IF THE SUM OF TWO NUMBERS IS GIVEN TOGETHER WITH THE RATIO OF THEIR PRODUCT TO THIS SUM, THEN EACH OF THEM CAN BE FOUND.
Let the sum be $x + y$, the product of x and y be b, and the ratio of xy and $x + y$ be given. Quadruple b to get d whose ratio to $x + y$ is also known. Call this e. Now take three numbers in proportion, f, g, and h, the greatest of which is h. Let the difference of h and f be k. Now the ratio of h to k is the ratio of something to e, and that is the known l. So add l to e and m. Now therefore l is to e as h is to k; hence, l is to m as h is to f. As square is to square, so is l times m a square. Now the square of l is n, which equals the product of l and m and l and e. But l times e is d. Hence, l times m is

the excess of n over d, which is z. Therefore z is a square whose root is t. If, therefore, l is divided in two whose difference is t, that will be as x and y. For the product of one times the other taken four times is d. The square of the difference is added. But if the three are taken for g, f, and h, the fourth must be l.

For example, let the product of the numbers be twice their sum. Hence, four times the product is eight times the sum. Let there be three numbers in proportion, 1, 2, and 4. The difference of four and one is three. Now 4 is to 3 as $10\frac{2}{3}$ is to 8. Therefore multiply $10\frac{2}{3}$ by $2\frac{2}{3}$ to get $28\frac{4}{9}$, whose root is $5\frac{1}{3}$. Subtract this from $10\frac{2}{3}$, halve the remainder, and get $2\frac{2}{3}$, which is one of the numbers. The other is 8, and the whole number was $10\frac{2}{3}$, because the product of $2\frac{2}{3}$ and 8 is twice $10\frac{2}{3}$, as was given. Note that if the proportional terms had been 1, 3, and 9, then the sum would have been 9 and the parts 6 and 3, for the same reason.

IV-26. IF THE RATIO OF THE SQUARES OF TWO NUMBERS TO THEIR SUM IS GIVEN TOGETHER WITH THE RATIO OF THEIR PRODUCT TO THEIR SUM, THEN EACH OF THEM CAN BE FOUND.

Let the numbers be x and y and their squares d and e and their product c. The following are given: the ratio of $d + e$ to $x + y$, of c to $x + y$, and of twice c to $x + y$. This latter is f. Hence, the ratio of $d + e + f$ to $x + y$ is known. But $x + y$ squared is $d + e + f$. Therefore $x + y$ is known. Now[134] c is the mean proportional between d and e. So $d + e$ to c is known and consequently d and c are found. But d to c is as x to y. Hence, since the ratio of x and y is known, x and y can be found.

For example, let the ratio of the sum of the squares of x and y to the sum of x and y be $3\frac{4}{7}$, and let the product of x and y to their sum be $1\frac{5}{7}$. Therefore twice the product and the two squares is seven times the square of the sum. Hence, the sum will be 7, the two squares 25, and the two numbers 3 and 4.

IV-27. IF TWO KNOWN NUMBERS ARE EACH ADDED TO A GIVEN MULTIPLE OF A NUMBER AND THE PRODUCT OF THE SUMS IS GIVEN, THEN THE NUMBER CAN BE FOUND.

Let d and e be added to the known ratios of b and x, and of c and x. Let the product of $b + d$ and $c + e$ be $f + g + h + l$, so that b times c is f, b times e is g, c times d is h, and d times e is l. Since b and c are to a, and h is to b and c, so will g and h be given with respect to a, and f given with respect to the square of a, which shall be z. Now l is known. Subtract this from $f + g + h + l$ to leave $f + g + h$ known. This consists of one given with respect to the root a and of another given with respect to the square z. Hence f, h, and a are all found.

For example, let one be added to a third of a root and to a fourth of a root. Let the product of the sums be 20. The third times the fourth makes a twelfth of the square, and the third times one and the fourth times one make a third and a fourth of the root, and one is one. Subtract this (one) from 20 to obtain 19, which is $\frac{7}{12}$ of the root and $\frac{1}{12}$ of the square. Therefore the square and seven roots equals 228. So add the square of $3\frac{1}{2}$ twice to 228 to get $240\frac{1}{4}$. The root of this is $15\frac{1}{2}$. $3\frac{1}{2}$ substracted from this leaves 12, and that is the number.

IV-28. ITEM, IF A GIVEN NUMBER IS ADDED TO ANOTHER NUMBER AND THE SUM IS MULTIPLIED BY WHAT REMAINS TO MAKE A GIVEN NUMBER, THEN THE NUMBER CAN BE FOUND.

Let b be the given number that is added to x; and let the product of $x + b$ and c be given as $e + f$. The product of c and b is e, and this is given with respect to z.[135] But the product of c and x is f, and this is given with respect to a. Hence, a number given with respect to the root with a number given with respect to the square makes the given number. Therefore a is given.

For example, let a fourth of a root multiplied by half a root increased by 1 equal 10. Now the half multiplied by the fourth makes an eighth of a square, and one times the fourth makes a fourth of the root. Hence, 10 equals a fourth of a root and an eighth of a square. Therefore the root is 8.

IV-29. IF TWO GIVEN NUMBERS ARE SUBTRACTED FROM TWO OTHER NUMBERS GIVEN IN RATIOS TO A THIRD NUMBER AND THE PRODUCT OF THE REMAINDERS IS GIVEN, THEN THAT NUMBER CAN BE FOUND.[136]

Let $b + c$ and $d + e$ be given with respect to a.[137] Now c and e are also given. Let the product of b and d be g, which is given. Let $b + c$ times e be h, and let $d + e$ times c be k. Hence, the product of $b + c$ and $d + e$ is l, and the product of c and e is m. Now, it follows that h is the sum of the products of b and e and of c and e. But b times e and b time d equals

b times $d + e$. Therefore $g + h$ equals m and the product of $d + e$ and b. But $d + e$ times b and c is l. Therefore $g + h + k$ equals $l + m$. Now the difference of g and m is n. Since they are given, n is known. But $h + k$ is given with respect to a and l is given with respect to e.[138] These are always equal to the difference, g. Therefore, either $h + k$ with n is l, or $n + l$ will be $h + k$. But both of these are known.

For example, let one number be 2 roots and the other 3 roots. Subtract 6 from the 3 roots and 4 from the 2 roots, and let the product of the remainders be 150. Now 2 times 3 makes 6 squares and 6 times 4 is 24. Similarly 3 times 4 makes 12 roots and 2 times 6 makes 12 roots. Therefore 6 squares and 24 equals 24 roots and 150. Subtracting 24 from 150 leaves 126 and 24 roots equal to 6 squares. Hence, one square equals 4 roots and 21. Now add the square of half of 4, which is 4, to 21 to make it 25. Its root is 5. To this add that half of 4 to obtain 7, which is the desired root.

IV-30.   IF A GIVEN NUMBER IS SUBTRACTED FROM SOME NUMBER AND THE PRODUCT OF THE REMAINDER AND WHAT REMAINS IS KNOWN, THEN THE NUMBER CAN BE FOUND.

Let $b + c$ and d be given with respect to a, and let the product of b and d be given as e. c is also given. Now the product of $b + c$ and d is $e + f$, and thus f equals the product of c and d. $e + f$ is given with respect to z and f is given with respect to a. Since a number is given with respect to a square, it follows from the given number and the number given with respect to the root that the root can be found.

For example, let x be twice y and z be $\frac{2}{5}$ of the same y.[139] From x subtract 4 and multiply the remainder by z to obtain 12. Now if all of x is multiplied by z, then you have $\frac{4}{5}$ of a square, and 4 multiplied by z yields $\frac{8}{5}$ of the root. Therefore $\frac{4}{5}$ of the square equals $\frac{8}{5}$ of the root and 12. Therefore, according to the previous operation the root is 5 and the square is 25.

IV-31.   IF THERE ARE TWO KNOWN MULTIPLES OF A NUMBER SUCH THAT A KNOWN NUMBER IS SUBTRACTED FROM ONE, ANOTHER ADDED TO THE OTHER, AND THE PRODUCT OF THE SUM AND DIFFERENCE IS GIVEN, THEN THE NUMBER CAN BE FOUND.

Let $b + c$ and d be given with respect to a; let c be given; let the known e be added to d; let the product of b and $d + e$ be f, which is given; and let the product of $b + c$ and $d + e$ be $g + h$, where g is the product of $b + c$ and d. Call the product of $b + c$ and e, h, and the product of $d + e$ and c, $l + m$, so that h equals $f + l + m$. Let l be given with respect to a, and let m be just given. Now the difference of l and h is that of the known t and a. Therefore, either t with the known $f + m$ is as g given with respect to z, or, $f + m$ is known equal to g given with respect to a, and g is known with respect to z. Whichever way it is, a is always known.

For example, let x be $\frac{3}{2}$ of y and z $\frac{1}{2}$ of y.[140] From x subtract one, to z add 3, and let the product of these be 25. Now x times z and 3 makes $\frac{3}{4}$ of a square and $4\frac{1}{2}$ roots. One times z and 3 yields $\frac{1}{2}$ root and 3. Hence, $28\frac{1}{2}$ equals $\frac{3}{4}$ of a square and four roots and one half. Consequently 28 equals $\frac{3}{4}$ of a square and four roots. And so, as before, the root is 4.

IV-32.   IF GIVEN NUMBERS ARE ADDED TO OR SUBTRACTED FROM THE SAME NUMBERS AND THE PRODUCT OF WHAT REMAINS IS KNOWN IN A RATIO WITH A ROOT OR A SQUARE, THEN THE ROOT AND THE SQUARE CAN BE FOUND.

If the numbers are added: if the product is given with respect to the square, then there remains a number given with respect to the square consisting of a given number and a number given with respect to the root after the product of the constants has been subtracted. If the product is given with respect to the root, then a number given with respect to the root remains consisting of a given number and a number given with respect to the square, after the product of the addends has been subtracted. If the numbers are subtracted: if the product is given with respect to the square, then as before the result follows in the manner of the demonstration for proposition 29 above. If the product is given with respect to the root, the result follows similarly.

For example, let x be $\frac{4}{3}$ z and let y be twice z. To x add 2 and to y add 3; and let their product be 6 squares. From this subtract what comes from multiplying x alone by y alone; the remainder is $3\frac{1}{3}$ of a square, which is equal to 8 roots and 6. And so the root is found to be 3 and the square is 9.

IV-33.   LIKEWISE, IF A GIVEN NUMBER IS ADDED OR SUBTRACTED FROM ONE NUMBER AND THIS IS MULTIPLIED BY THE OTHER NUMBER TO PRODUCE A NUMBER KNOWN WITH RESPECT TO A SQUARE OR A ROOT, THEN THE ROOT WITH THE SQUARE CAN BE FOUND.

Whether the given number is added or subtracted, if the product is given with respect to a square (for the product of two numbers makes the ratio of one to the other known), then the product of a known number and the other number is given with respect to it. But let it be given with respect to the root. Therefore the root is known with respect to the square. And so, if the given number is added and the product is given with respect to the root, there will be the product of the other number and the root. A similar thing follows if the given number is subtracted. Hence, the ratio of the root to the square is known, and the root is found.

For example, let x be $\frac{4}{3}$ of z and y be $\frac{1}{3}$ of z. Add 2 to x and let the product of the sum and y be $\frac{2}{3}$ of a square. But $\frac{1}{3}$ of a root times $\frac{4}{3}$ of a root makes $\frac{4}{9}$ of a square, and $\frac{1}{3}$ of a root times 2 makes $\frac{2}{3}$ of a root, which is equal to $\frac{2}{9}$ of the square. Hence, the root will be $\frac{1}{3}$ of the square. Therefore the root, or z, will be 3.

IV-34.   IF ONE OF THE KNOWN MULTIPLES IS ADDED TO A GIVEN NUMBER AND FROM THE OTHER IS SUBTRACTED A GIVEN NUMBER SUCH THAT THE PRODUCT OF THE SUM AND DIFFERENCE MAKES A KNOWN NUMBER OF MULTIPLES OF EITHER THE ROOT OR THE SQUARE, THEN THE SQUARE WITH ITS ROOT CAN BE FOUND.

If the product is given with respect to the square: then the product of the sum and the difference equals a number given with respect to the root, and the number will be found; or, the product of the sum and the difference equals a number given with respect to the square, and the number given with respect to the root or to the square; or, the given with respect to the square and another given with respect to the root. Subtract the smaller from the larger. What remains will be equal to the number given wih respect to the root or to the square; or, the given number will be equal to the number given with respect to the root or to the square; or, the number given equal to the number given with respect to the root will be equal to a given number and a number given with respect to the square; or, the number given with respect to the square will be equal to the given number and a number given with respect to the root. Whichever happens, the root and the square are found. If the root is given with respect to the square: then for the same reason the product given with respect to the root with a given number will equal a number given with respect to the root and a number given with respect to the square. After the subtractions, either there will be a number given equal to a number given with respect to a square with a number given with respect to the root, or there will be a number given with respect to the square equal to a given number and a number given with respect to the root. Thus the root will be found.

For example, let one number be twice the root and another thrice it. To the former add 4, from the latter subtract 6. Let the product be $3\frac{1}{3}$ of a square. Now the first number times 6 is 12 roots and 4 times 6 is 24. 3 roots times 2 roots equals six squares, and 3 times 4 is twelve. The result is 6 squares and 12 roots equals $3\frac{1}{3}$ squares and 12 roots and 24. Subtracting from both sides the 12 roots and taking $3\frac{1}{3}$ squares from 6, what remains is $2\frac{2}{3}$ squares and 24. The square therefore is 9. The root is 3, and so on as before.

IV-35. IF A NUMBER GIVEN WITH RESPECT TO A SQUARE EQUALS A NUMBER GIVEN WITH RESPECT TO A ROOT, THEN THE ROOT CAN BE FOUND.

Let x be the root, b be the square, e be a number given with respect to the square, the square of e be g given with respect to x, and d be the square of b. Now between b and d lies a mean proportional to the ratio of b and x; call this c. Since x squared is b, and x times b is c, then x times c is d. Since the ratio of g and d is given, the ratio of x to d is given; hence, c is found to be the cube of x. Take the cube root, therefore, to obtain x.

For example, let the square of half a square be equal to 54 roots. Now the square of the square equals $\frac{1}{2}$ squared squares and four times 54 is 216. The cube root of this is 6, the root of the square.

The End of This Book and of
the Entire Book of Given Numbers

Notes to the Translation

1. That is, the ratio of proportionality.

2. *Cf.* al-Khwārizmī, *Liber algebre*, iii; *Kitāb fī al-jābr wa'l muqābala*, no. 5; Fibonacci, *Liber abaci*, 457. This is the first of at least five problems that can be matched with similar problems in *Liber mensurationum* of abū Bekr (H. L. L. Busard, "L'Algebre au moyen age: le 'Liber Mensurationum' d'Abū Bekr," *Journal des Savants*, Apr.–June 1968, pp. 65–124); namely, I-3 and no. 25, I-4 and no. 28, I-5 and no. 26, I-15 and no. 30, IV-10 and no. 27. The parallelism between problems, however, does not confirm any influence of abū Bekr on Jordanus, particularly since the numbers and methods of solution differ. *Cf.* Euclid, *Data*, 85.

3. *Cf.* al-Khwārizmī, *Liber algebre*, 107; Fibonacci, *Liber abaci*, 411; abū Kāmil, *Kitāb*, no. 8.1.

4. Jordanus neglects to mention that this is equal to $2a^2 - b$.

5. By I-1.   6. In I-4; cf. Euclid, *Data*, 84.

7. B has an insert that lets the reader know why this is true, "Now by proposition 17 of Jordanus' *Arithmetica*." The proposition here is I-17. "The square of the sum of the parts of a number is equal to the sum of four times their parts and the square of their differences."

8. By I-1.   9. By I-1.   10. By I-5.

11. From Jordanus' *Arithmetica*, I-18.

12. The following note was added to D: "Euclid II, 9 on numbers makes it obvious that if a number is squared and also multiplied by another, such as a squared and ab, the square of the whole product with the square of b is a square number whose root is b and twice a. This is obvious from Jordanus' *Arithmetica* I, 9 and 17, which justifies the mechanics and truth of this work as well as of many of the following theorems." Another gloss at the same place in D says, "from the seventh corollary."

13. By I-7.

14. Jordanus adds and subtracts $(x - y)^2$ to the right side of the equation twice to obtain the critical $3(x - y)^2$. This may be the first instance in medieval mathematics of a modern technique used in proof of algebraic theorems.

15. By I-1.   16. In I-12, the "thrice the square of the difference."

17. See al-Khwārizmī, *Liber algebre*, iii; abū Kāmil, *Kitāb*, no. 7.

18. See al-Khwārizmī, *Liber algebre*, 111.

19. Note the neat factoring which, strangely, does not enter into the solution of the example for the theorem.

20. By I-7; then by I-1 for x and y.

21. "... unique," i.e., the student cannot determine which of the two possibilities, $(x - y) = 2$ or $(x - y) = 1$, is correct; both are admissible yet both can result in error, as the example demonstrates. The error is that there is not a unique pair of values for x and y.

22. Jordanus gives no hint of how he solved this equation for two values of $(x - y)$. Notice that he calls this "an error."

23. See Fibonacci, *Liber abaci*, 428.   24. By I-7 and I-1.

25. By I-1.

26. See al-Khwārizmī, *Liber algebre*, 105; abū Kāmil, *Kitāb*, no. 3.

27. See abū Kāmil, *Kitāb*, no. 8; Fibonacci, *Liber abaci*, 434 and 441.

28. By I-1.

29. Jordanus errs here. The result is not the product of the parts but the sum of their squares. The problem is eventually solved by I-4 and I-1.

30. The reference is to Jordanus' *Arithmetica*, I-17; see I-4 above.

31. See Fibonacci, *Liber abaci*, 440 and 454.

32. By I-3.   33. See Fibonacci, *Liber abaci*, 457.

34. See Fibonacci, *Liber abaci*, 458.

35. Which is xy. Since $x + y$ is given, the conclusion follows.

36. By I-3.

37. The statement of the theorem presupposes the information in I-28,β.

38. In the demonstration Jordanus works from $\dfrac{a + b}{e + c} = \dfrac{a}{e}$. In the example, he uses $\dfrac{b}{c}$ for $\dfrac{a}{e}$.

39. In I-22,α.   40. By I-19.

41. This is I-23*, unique to the Beta-2 family.

42. See Fibonacci, *Liber abaci*, 419.

43. Jordanus sets $e = \frac{1}{2}$ and does not prove the theorem for any e, although a simple substitution of e for $\frac{1}{2}$ proves the general theorem.

44. d is found by I-7, and x by simple subtraction, as in the example.

45. It is difficult to appreciate the point Jordanus is making here since, after the initial division by c, this theorem is reduced to the third case of I-24.

46. By I-22.

47. One may wonder why Jordanus did not first multiply 2 by 4, then both sides of the equation by 8.

48. By I-22,A, then by I-3 and I-1.

49. Again, Jordanus is using the quotient from I-26.

50. Note Jordanus' awareness of the associative law for multiplication.

51. By I-23.

52. Note that Jordanus did not work the example in the way that he proved the theorem.

53. See Fibonacci, *Liber abaci*, 440.

54. See al-Khwārizmī, *Liber algebre*, 112; abū Kāmil, *Kitāb*, no. 4; Fibonacci, *Liber abaci*, 410.

55. Jordanus adds $xy + x^2$ to both sides of the first equation to reach the third, which is then solved by I-7.

56. By I-7.

57. Jordanus might have inserted the word *approximated* here, as the theorem suggests.

58. See Euclid, *Data*, 2.

59. This is an early example of continued unitary fractions in the Egyptian fashion. Compare with M. Duton and R. E. Grim, "Fibonacci on Egyptian Fractions," *The Fibonacci Quarterly*, 1966, *4*:339–54.

60. See Euclid, *Data*, 5.   61. See Euclid, *Data*, 7.

62. Note that Jordanus gives the sum as $1 + \frac{1}{5} + \frac{1}{15}$, in the Egyptian fashion.

63. By II-11.

64. The proposition states that these ratios are given to start with. The contradiction is obvious.

65. The example makes sense only if $y = \frac{1}{2}x$, with $x = x + a$ in the demonstration.

66. By I-7.   67. Which equals $1\frac{1}{8}$.

68. See Fibonacci, *Liber abaci*, 416.   69. As in II-18.

70. See Euclid, *Data*, 7.

71. Jordanus' parenthetical expression tends to generalize the theorem.

72. See Fibonacci, *Liber abaci*, 454.

73. The last sentence is a check for the solution.

74. Note Jordanus' awareness of the associative and commutative laws for addition.

75. G. Wertheim, "Ueber die Lösung einiger Aufgaben im *Tractatus de numeris datis* des Jordanus Nemorarius," *Bibliotheca Mathematica*, 1900, Ser. 3, *1*:418–20, offers a clear, modern symbolic translation of the example.

76. See Fibonacci, *Liber abaci*, 327.

77. What makes this so confusing is that Jordanus does not tell his reader that these ks, js, ms, and ns are multiples of e. The example is clear.

78. This is the first example in *De datis* of clearing fractions from an equation by multiplying through by the denominator.

79. What follows is an excellent example of the simultaneous solution of several equations. The modern reader will keep his sense of

direction through this mathematical maze if he remembers two things: the words *first, second, third,* and *fourth* refer exclusively to the four unknowns; and the other numbers, both integral and fractional, are the coefficients of the unknowns. If he will write out the four equations in order, expand the parentheses, and subtract the second equation from the first, he will see what lies in the next two sentences of the translation.

80. Wertheim (n. 75 above), 417, shows cogently that this example and solution is in all respects parallel to II-28 in the treatise on algebra, *al-Fakhrī* by al-Karajī, which in turn parallels I-25 in Diophantus' *Arithmetica*. What Heath remarks about the early history of manuscripts of *Arithmetica* led me to disregard it as a possible source for Jordanus. (see T. L. Heath, *Diophantus of Alexandria* [Cambridge, 1910], 14–15). Furthermore, had Jordanus had it at hand, I do not think he would have resisted incorporating more than the half-dozen theorems I found in *De datis* that are the same as those in *Arithmetica*.

81. That is, $y_1 + \frac{1}{3}(12 - y_1) = 6$.

82. This is a fairly literal translation since the meaning is hidden in the maze of letters.

83. See Fibonacci, *Liber abaci*, 410.

84. My addition for some clarity.

85. My addition for some clarity.

86. Since the first number was zero, Jordanus did not count it.

87. My addition for some clarity.

88. The equations are $w_1 + 3(28 - w_1) = 84 \rightarrow w_1 = 0$, and so on. Then, $w_2 = w_1 + 3\frac{1}{9}$, and so on. These values substituted into the original equations yield $87\frac{1}{9}$, instead of 28. Hence, the correct ratio is $28 : 87\frac{1}{9}$ or $9 : 28$ by which the second values are multiplied to produce $w = 1$, $x = 2$, $y = 3$ and $z = 4$. *Cf.* Curtze, *Commentar*, 63.

89. Note that the adjective *proportionalis* corresponds to the noun *proportionalitas* (proportion) and not to *proportio* (ratio).

90. Literally, "hence the first contains the second once and a half."

91. By I-3.

92. Note that the proof of the theorem suggests I-3 as the route toward the desired numbers, while the example uses I-1.

93. By I-3 and I-1.   94. By I-7.   95. By I-7.

96. This implies a quartic equation, as the algebra indicates.

97. By III-4.   98. By I-5.   99. Jordanus uses d as a parameter.

100. The word *duplus* generally means the *square* of a number, as translated here. But the precise point Jordanus makes in this sentence escapes me. Apparently he does not use this statement further in the development of the proof.

101. By I-3.

102. At this time in the history of mathematics, negative roots for equations were ignored; as a result, mathematicians were often disturbed when their equations did not produce answers immediately. This may account for Jordanus' comment, "it is not clear which is which."

103. By I-5.

104. Jordanus uses the identity element for multiplication, $\frac{x}{x} = 1$ and $\frac{y}{y} = 1$, twice here.

105. Jordanus uses the commutative law for multiplication here:
$$\left(\frac{a}{y}\right)\left(\frac{x}{d}\right) = \left(\frac{a}{d}\right)\left(\frac{x}{y}\right).$$

106. Jordanus manipulates ratios of ratios. Hence, to keep the translation simple, the colon is used for the interior ratios.

107. By III-15 or by I-7, as in the example.

108. The importance of the combination escapes me, since the sum of the terms may be had without going through this step.

109. Jordanus uses the word *differentia*. As Curtze points out, in the Middle Ages "to subtract one ratio from another" meant to take their quotient; similarly, "to add one ratio to another" meant "to take their product" (see his "Commentar zu dem "Tractatus *De numeris datis*' des Jordanus Nemorarius," *Zeitschrift für Mathematik und Physik, Hist.-lit. Abtg.* 1891, *36*:95). Jordanus uses the word *dividatur* in his example instead of the word *differentia*: this confirms Curtze's observation.

110. B has as part of the text, "Note that the divisor is the number by which the division is done and the dividing number is called the quotient."

111. The example assumes $x > y$.   112. By I-1.   113. By I-17.

114. See Euclid, *Elements*, II,5.

115. Note the rule $(a + b)(a - b) = a^2 - b^2$.

116. Here is one of the rare occasions when Jordanus mixed algebraic and geometric thought.

117. See al-Khwārizmī, *Liber algebre*, 71; Euclid, *Elements*, II,11.

118. Jordanus has completed the square but he (the scribe?) set $c = d$.

119. The Cartesian convention was not used here, in order to illustrate the difficulty the reader may experience in distinguishing variables from constants.

120. See al-Khwārizmī,   121. Ibid., 77.

122. That is, 4, the coefficient of x.   123. By IV-8.

124. As in IV-10, Jordanus ignores the negative root.

125. By I-3.   126. Note the indeterminate solution.

127. As in IV-14.   128. As in IV-14.

129. This is an unusual example. Ordinarily, Jordanus brings the reader to one of the unknowns in the demonstration. But to leave the example without finding all the numbers is not his usual method. Perhaps this was his way of letting the reader know that he must begin to do more for himself.

130. The word *numbers*, which appears here and in the statements of the next four propositions, is some form of the Latin *latus*. The significance of this is discussed in the Introduction.

131. This last sentence is an unexpected intrusion. The problem has been solved; the answers are 3 and 6. Suddenly, this remark appears. Dresden C 80, one of the digests discussed in the Introduction, has almost the same remark: "But it would be better after the fashion of propositions 1 and 3 of Book I to subtract. . . . Hence by I-1, the numbers are found, as before." The addition is probably a marginal gloss that became incorporated into the text. I suspect that the occasion for the gloss was the next proposition, in which the example is solved after the manner stated by the glossist.

132. See abū Kāmil, *Kitāb*, 7.

133. This is the first time a cubic equation is used in *De datis*.

134. Note a second solution, not used in the example.

135. The z here and the a in the next sentence make the coefficients of the unknown rational numbers.

136. See Euclid, *Elements*, VI, 28 and 29.

137. The wording suggests that the ratios $(b + c):a$ and $(d + e):a$ are given; the sense of the demonstration denies this. Rather, the idea, in modern symbols, is:

$$b:x = p_1 \qquad\qquad\qquad d:x = p_2$$
$$b = p_1 x \qquad\qquad\qquad\quad d = p_2 x$$
$$b + c = p_1 x + c \qquad\qquad d + e = p_2 x + e$$

138. This sentence and the next two are confusing.

139. x, y, and z do not appear in the text; their use, however, simplifies the translation.

140. See n. 139, above.

The Symbolic
Translation

Book-Proposition	Equations from Hypothesis	Reduction to Canonical Form	Final Resolution by
I-1	$x + y = a,\ x - y = b$	$y = \frac{1}{2}(a - b)$	
I-2	$w + x + y + z = a,\ w - x = b,$ $x - y = c,\ y - z = d$	$z = [a - (b + 2c + 3d)]$	
I-3	$x + y = a,\ xy = b$	$x - y = (a^2 - 4b)^{\frac{1}{2}}$	I-1
I-4	$x + y = a,\ x^2 + y^2 = b$	$x - y = (2b - a^2)^{\frac{1}{2}}$	I-1
I-5	$x - y = a,\ xy = b$	$x + y = (4b + a^2)^{\frac{1}{2}}$	I-1
I-6	$x - y = a,\ x^2 + y^2 = b$	$x + y = (2b - a^2)^{\frac{1}{2}}$	I-1
I-7	$x + b = y,\ x^2 + bx = a$	$x = \frac{1}{2}[(b^2 + 4a)^{\frac{1}{2}} - b]$	
I-8	$x + y = a,\ (x+y)(x-y) + y^2 = b$	$x = b^{\frac{1}{2}}$	
I-9	$x + y = a,\ (x+y)(x-y) + x^2 = b$	$x = \frac{1}{2}[(8a^2 + 4b)^{\frac{1}{2}} - 2a]$	
I-10	$x + y = a,$ $x^2 + y^2 + (x + y)(x - y) = b$	$x = (\frac{1}{2}b)^{\frac{1}{2}}$	
I-11	$x + y = a,$ $(x + y)(x - y) + xy = b$	$y^2 + ay = a^2 - b$	I-7
I-12	$x + y = a,\ x^2 + y^2 + (x-y)^2 = b$	$x - y = [\frac{1}{3}(2b - a^2)]^{\frac{1}{2}}$	I-1
I-13	$x + y = a,\ xy + (x - y)^2 = b$	$x - y = [\frac{1}{3}(4b - a^2)]$	I-1
I-14	$x + y = a,\ x^2 - y^2 = b$	$y = \frac{1}{2}(a^2 - b)/a$	
I-15	$x + y = a,\ x^2 + y^2 + x - y = b$	$(x - y)^2 + 2(x - y) = 2b - a^2$	I-7&1
I-16	$x + y = a,\ xy + x - y = b$	$a^2 < 2b,\ 2b - a^2 = 4,\ x - y = 2$	I-1
		$2b - a^2 = 3,$ $x - y = 2 \pm 1$	I-1
		$a^2 = 2b,\ x - y = 4$	I-1
		$a^2 > 2b,\ x - y = 2 \pm (4 - 4b + a^2)^{\frac{1}{2}}$	I-1

Book-Proposition	Equations from Hypothesis	Reduction to Canonical Form	Final Resolution by
I-17	$x + y = a$, $xy/(x - y) = b$	$x - y = (4b^2 + a^2)^{\frac{1}{2}} - 2b$	I-1
I-18	$x + y = a$, $(x^2 + y^2)/(x - y) = b$	$x - y = b - (b^2 - a^2)^{\frac{1}{2}}$	I-1
I-19	$x + y = a$, $x/y = b$	$y = a/(b + 1)$	
I-20	$x + y = a$, $x/y + y/x = b$	$x - y = a[1 - 4/(b + 2)]^{\frac{1}{2}}$	I-1
I-21,α	$x + y = a$, $c/x + c/y = b$	$xy = ac/b$	I-3
I-21,β	$x + y = a$, $(cx)y = b$	$xy = b/c$	I-3
I-22,α	$x + y = a$, $c/xy = b$	$xy = c/b$	I-3
I-22,β	$x + y = a$, $cx/y = b$	$y = ca/(b + c)$	
I-23	$x + y = a$, $cy/cx = b$	$x = a/(b + 1)$	
I-23*	$x + y = a$, $x/cy = b$	$x = a/(bc + 1)$	
I-24	$x + y = a$, $x + x/ny = b$	$a > b$, $z(z + g) = b/n$, $z = x/ny$ and $g = a - b + 1/n$	I-7
		$a < b$, $y(y + g) = a/n$	I-7
		$a = b$, $y(y + 1/n) = b/n$	I-7
I-25	$x + y = a$, $cx + cx/ny = b$	As in I-24, with b divided everywhere by c.	
I-26	$x + y = a$, $x/c + y/d = b$	$x = c(a - bd)/(c - d)$	
I-27	$x + y = a$, $(x/c)(y/d) = b$	$xy = bcd$	I-3
I-28,α	$x + y = a$, $(x/c)/(y/d) = b$	$x/y = bc/d$	I-23

Book-Proposition	Equations from Hypothesis	Reduction to Canonical Form	Final Resolution by
I-28,β	$x + y = a,\ c/x + c/y = b$	$xy = ca/b$	I-3
I-29	$x + y = a,\ ay = x^2$	$x^2 + ax = a^2$	I-7
II-1	$a:x = b:c$	$x = ac/b$	
II-2	$x:a = b$	$x = ab$	
II-3	$a:b = c$	$b:a = 1/c$	
II-4	$x + y = a,\ (x + y):x = b,\ y:x = c$	$y:x = b - 1,$ $(x + y):x = c + 1$	
II-5	$x + y = a,\ (x + y):x = b$	$a:y = b/(b - 1)$	
II-6	$x + y = a,\ x:y = b$	$y = a/(b + 1)$	
II-7	$x:y = a,\ y:z = b$	$x:z = ab$	
II-8	$x:n = a,\ y:n = b,\ \ldots,\ z:n = c$	$(x + y + \cdots + z): n = a + b + \cdots + c$	
II-9	$n:x = a,\ n:y = b,\ \ldots,\ n:z = c$	$(x + y + \cdots + z): n = 1/a + 1/b + \cdots + 1/c$	
II-10	$x + y = a,\ w + z = b,\ x:z = c,$ $y:w = d$	$x = c(bd - a)/ (d - c)$	
II-11	$x:y = a,\ (x + b):(y - c) = d$	$y = (b + cd)/ (a - d)$	
II-12,α	$(x + a):(y - b) = c,\ (y + d): (x - e) = f$	$y = [d + (a + bc + e)f]/(cf - 1)$	
II-12,β	$x:y = a,\ (x - b):(y - c) = d$	$y = (b - dc)/ (a - b)$	
II-13,α	$a - x = w,\ b - y = z,\ x:y = c,$ $w - z = d$	$x = cd(c - 1)$	
II-13,β	$x:y = a,\ (x - b)/(y - y/c) = d$	$y = bc/(ac - cd + d)$	

Book-Proposition	Equations from Hypothesis	Reduction to Canonical Form	Final Resolution by
II-14,α	$x:y = a,\ b - x = w,\ c - y = z,$ $wz = d$	$d = (b - ya)(c - y)$	I-7
II-14β	$x:y = a,\ (x - b)/c + y = d$	$y = (cd + b)/(a + c)$	
II-15	$x = w + y + \cdots + a,\ x:w = b,$ $x:y = c,\ \ldots$	$x = a/[1 - (b +$ $c + \cdots)]$	
II-16	$x = w + y + \ldots + z,\ (z + c):$ $x = f,\ w:x = a,\ y:x = b,\ \ldots$	$x = c/(a + b +$ $\cdots + f - 1)$	
II-17	$x = w + y + \cdots + z,\ w + x:$ $b = a,\ y:x = c,\ z:x = d,\ \ldots$	$x = a/(1 - c -$ $d + 1/b)$	
II-18	$x + y + \cdots + z = a,\ x:y =$ $b,\ \ldots,\ y:z = c$	$x = a/(1 + 1/b +$ $1/bc)$	
II-19	$x + y + \cdots + z = a,\ (x + b):$ $(y + c) = e,\ (y + c):$ $(z + d) = f$	$x = ef(a + b + c +$ $d)/(1 + f +$ $ef) - b$	
II-20	$(x + a):y = d,\ (y + b):z = e,$ $(z + c):x = f$	$x = (a + bd + cde)/$ $(def - 1)$	
II-21	$x + y = a,\ x + y:b = c$	$y = b(a - c)/(b - 1)$	
II-22	$x + y = a,\ (x + c):y = b$	$y = (a + c)/(1 + b)$	
II-23	$x + y + z = a,\ x:(y + z) = b,$ $y:(z + x) = c$	$x = (bc + b)z/$ $(1 - bc)$ $y = (bc + c)z/$ $(1 - bc)$	
II-24	$x + y + z = a,\ (x + b):(y + z) =$ $c,\ (y + b):(z + x) = d,$ $(z + b):(x + y) = e$	Solve by recursive substitution for one equation in one unknown	
II-25	$x + by = a,\ y + cz = a,$ $z + dx = a$	As in II-24	
II-26	$x + a(y + z) = d,\ y + b(z + x) =$ $d,\ z + c(x + y) = d$	As in II-24	

Book-Proposition	Equations from Hypothesis	Reduction to Canonical Form	Final Resolution by
II-27	Same as II-26	Solve by *Regula falsae positionis*	
II-28	Same as II-26	Another solution by *Regula falsae positionis*	
III-1	$a:x:b$	$x = (ab)^{\frac{1}{2}}$	
III-2	$x:a:b$	$x = a^2/b$	
III-3	$x:y:z,\ x:y = a$	$x:z = a^2$	
III-4	$x:y:z,\ x:z = a$	$x:y = a^{\frac{1}{2}}$	
III-5	$x:b:y,\ x + y = a$	$xy = b^2$	I-3
III-6	$x:y:z,\ (x + z):y = a$	$(x/y):(z/y) = 1$	I-3
III-7	$a:x:y,\ x + y = b$	$y^2 + ay = ab$	I-7
III-8	$x:y:z,\ (x + y):z = a$	$(y/z)^2 + y/z = a$	I-7
III-9	$x:y:b,\ x + 2y = a$	$y = (b^2 + ab)^{\frac{1}{2}} - b$	
III-10	$x:y:z,\ x + y + z = a,\ x:z = b$	$x:y = b^{\frac{1}{2}} = y:z$	III-4
III-11	$x:y:z,\ x + y + z = a,\ x - z = b$	$y^2 + \frac{2}{3}ay = \frac{1}{3}(a^2 - b^2)$	I-7
III-12	$x:y:z,\ x > z,\ x + z = a,$ $y + z = b$	$\left(\dfrac{x - z}{2}\right)^2 + \dfrac{x - z}{2}$ $= \dfrac{a^2}{4} + \dfrac{a}{2} - b$	I-7 & I
III-13	$x:y:z,\ x > z,\ x + y = a,$ $x + z = b$	$y = \dfrac{1}{2}\left\{ a - \dfrac{b^2}{2} \pm \left[\dfrac{b^2}{4} \right.\right.$ $\left. - \left\{ \left(a - \dfrac{b}{2} \right)^2 \right.\right.$ $\left.\left.\left. - \dfrac{b^2}{4} \right\} \right]^{\frac{1}{2}} \right\}$	

Book-Proposition	Equations from Hypothesis	Reduction to Canonical Form	Final Resolution by
III-14	$a:x = y:b,\ x + y = c$	$xy = ab$	I-3
III-15	$a:x = y:b,\ x - y = c$	$xy = ab$	I-5
III-16	$a:x = y:b,\ x:y = c$	$x = (abc)^{\frac{1}{3}}$	
III-17	$a:x = y:b,\ (x + a):y = c$	$x^2 + ax = abc$	I-7
III-18	$w:x = y:z,\ w + x = a,$ $y + z = b,\ w:z = c$	$x = abc/(bc + a)$	
III-19	$w:x = y:z,\ w + z = a,$ $x + y = b,\ w:y = c$	$w = \dfrac{c\left[a - \dfrac{a+b}{c+1}\right]}{(c-1)}$	
III-20	$w:x = y:z,\ w + x + y + z = a,$ $w - x = b,\ y - z = c$	$w = \dfrac{b(a + b + c)}{2(b + c)}$	
III-21	$w:x = y:z,\ w + x + y + z = a,$ $w - z = b,\ x - y = c$	$w = \dfrac{(b-c)(a+b+c)}{2(b+c)}$	
III-22	$u:v = v:w,\ x:y = y:z,$ $u:x = a,\ w:z = b$	$v:y = (ab)^{\frac{1}{2}}$	
III-23	$u:v = v:w,\ x:y = y:z,$ $u:x = a,\ v:y = b$	$w:z = b/a$	
IV-1	$w/y = u,\ x/z = v,\ w:x = a,$ $y:z = b$	$u:v = a:b$	
IV-2	$(w/y):(x/z) = a,\ y:z = b$	$w:x = ab$	
IV-3	$a/x = z,\ a/y = w,\ x - y = b,$ $z - w = c$	$x + y = (4ab/c + b^2)^{\frac{1}{2}}$	I-1
IV-4	$a/x - a/y = b,\ x + y = c$	$xy/(y - x) = a/b$	I-17
IV-5	$x:y = a,\ xy = b$	$x = (ab)^{\frac{1}{2}}$	
IV-6	$x:y = a,\ x^2 + y^2 = b$	$y = [b/(a^2 + 1)]^{\frac{1}{2}}$	
IV-7	$x:y = a,\ x^2 - y^2 = b$	$y = [b/(a^2 - 1)]^{\frac{1}{2}}$	

Book-Proposition	Equations from Hypothesis	Reduction to Canonical Form	Final Resolution by
IV-8	$x^2 + bx = c$	$x = \left(c + \dfrac{b^2}{4}\right)^{\frac{1}{2}} - \dfrac{b}{2}$	
IV-9	$x^2 + c = 2bx$	$x = b \pm (b^2 - c)^{\frac{1}{2}}$	
IV-10	$x^2 = 2bx + c$	$x = b + (b^2 + c)^{\frac{1}{2}}$	
IV-11	$x^2 : a + x : b = c$	$x^2 + ax/b = ac$	IV-8
IV-12	$x^2 : a + c = x : b$	$x^2 + ac = ax/b$	IV-9
IV-13	$x^2 : a = x : b + c$	$x^2 = ax/b + ac$	IV-10
IV-14	$(x + y) : z = a,\ xy : z^2 = b$	$x : z = \frac{1}{2}[a + (a^2 - 4b)^{\frac{1}{2}}]$ $y : z = \frac{1}{2}[a - (a^2 - 4b)^{\frac{1}{2}}]$	
IV-15	$(x + y) : z = a,\ (x^2 + y^2) : z^2 = b$	$x : z = \frac{1}{2}[a + (2b - a^2)^{\frac{1}{2}}]$ $y : z = \frac{1}{2}[a - (2b - a^2)^{\frac{1}{2}}]$	
IV-16	$(x - y) : z,\ xy : z^2 = b$	$x : z = \frac{1}{2}[(a^2 + 4b)^{\frac{1}{2}} + a]$ $y : z = \frac{1}{2}[(a^2 + 4b)^{\frac{1}{2}} - a]$	
IV-17	$(x - y) : z,\ (x^2 + y^2) : z^2 = b$	$x : y = \frac{1}{2}[(2b - a^2)^{\frac{1}{2}} + a]$ $y : z = \frac{1}{2}[(2b - a^2)^{\frac{1}{2}} - a]$	
IV-18	$x : a = z,\ y : z = b,\ xy : z = c$	$z = c/ab$	
IV-19	$xy = a,\ x^2 + y^2 = b$	$x^2 - y^2 = (b^2 - 4a^2)^{\frac{1}{2}}$	I-1
IV-20	$xy = a,\ x^2 - y^2 = b$	$x^2 + y^2 = (b^2 + 4a^2)^{\frac{1}{2}}$	I-1

Book-Propo-sition	Equations from Hypothesis	Reduction to Canonical Form	Final Reso-lution by
IV-21	$x + y = a$, $x^2y^2 = b$	$x - y = (a^2 - 4b^{\frac{1}{2}})^{\frac{1}{2}}$	I-1
IV-22	$x - y = a$, $x^2y^2 = b$	$x + y = (a^2 + 4b^{\frac{1}{2}})^{\frac{1}{2}}$	I-1
IV-23	$x - y = a$, $x^2 - y^2 = b$	$x + y = b/a$	I-1
IV-24	$x^2 : y = a$, $y^2 : x = b$	$x = (a^2b)^{\frac{1}{3}}$	
IV-25	$x + y = a$, $xy/(x + y) = b$	$xy = ab$	I-3
IV-26	$(x^2 + y^2):(x + y) = a$, $xy:(x + y) = b$	$x + y = a + 2b$	I-3
IV-27	$(x/b + d)(x/c + d) = a$	$x^2 + (b + c)dx = (a - d^2)bc$	IV-8
IV-28	$(x/a)(x/b + c) = d$	$x^2 + bcx = abd$	IV-8
IV-29	$(ax - c)(bx - d) = e$	$x^2 = (d/b + c/a)x + (e - cd)/ab$	IV-10
IV-30	$ax(bx - c) = e$	$x^2 = (c/b)x + e/ab$	IV-10
IV-31	$(ax + c)(bx - d) = e$	$x^2 + (c/a - d/b)x = (e + cd)/ab$	IV-8
IV-32	$(ax \pm c)(bx \pm d) = ex^2$ [or $= ex$]	By multiplying factors, collecting terms and reducing the coefficient of x^2 to one, the resultant equation is solved by IV-27, 29, or 31.	
IV-33	$ax(bx + c):x^2 = e$	$x = ac/(e - ab)$	
	$ax(bx + c):x = e$	$x = (e - ac)/ab$	
IV-34	$(ax + c)(bx - d) = ex^2$ [or $= ex$]	As in IV-32	
IV-35	$a:x^2 = b$, $a^2:x = c$	$x = (c/b^2)^{\frac{1}{3}}$	

Bibliography

Excluded from this list are the catalogues
describing the codices in which manuscript
copies of *De numeris datis* are found;
they are described with their respective
codices in the Introduction.

I. Primary Sources

ABRAHAM. *Liber augmenti et diminutionis vocatus numeratio divinationis*,
in G. Libri, *Histoire des sciences mathématiques en Italie*, I, pp. 304–372
(Paris, 1838).

(ABŪ BEKR). Busard, Hubert L. L. "L'Algèbre au moyen âge: le 'Liber
mensurationum' d'Abū Bekr," *Journal des Savants*, Apr.-June 1968,
pp. 65–124.

(ABŪ KĀMIL SŪJA' IBN ASLAM). Levey, Martin. *The Algebra of Abū Kāmil
in a Commentary by Mordecai Finzi*. Hebrew text, translation, and
commentary with special reference to the Arabic text. Madison: University of Wisconsin Press, 1966.

(AL-KHWĀRIZMĪ, MUHAMMAD IBN MUSA). Karpinski, Louis C. *Robert
of Chester's Latin Translation of the Algebra of al-Khowarizmi*. With
an introduction, critical notes, and an English version. New York:
Macmillan, 1915.

———.*Liber Maumeti filii Moysi alchoarismi de algebra et almuchabala*,
in G. Libri, *Histoire des sciences mathématiques en Italie*, I, pp. 253–297
(Paris, 1838).

CARDANO, GIROLAMO. *The Great Art or the Rules of Algebra*. Translated
by T. Richard Witmer. Cambridge: M.I.T. Press, 1968.

(DIOPHANTOS OF ALEXANDRIA). Heath, Sir Thomas Little. *Diophantos of Alexandria: A Study in the History of Greek Algebra.* 2d ed. Cambridge: University Press, 1910.

Euclid. *Euclidis Data cum commentario Marini Philosophi.* Translated into Latin by Claudius Hardy. Paris: Melchioris Mondiere, 1625.

———. *Euclidis Data cum commentario Marini et scholiis antiquis.* Edited and translated into Latin by H. Menge. Leipzig: B. G. Teubner, 1896.

———. *Euclidis Mergarensis, Geometricorum elementorum libri xv, Campani Falli transalpini in eosdem commentariorum libri xv.* Paris: Henricus Stephanus, 1516.

(———). Heath, Sir Thomas Little. *The Thirteen Books of Euclid's Elements.* Reprinted edition, 3 vols. New York: Dover, 1956.

JOHN OF SEVILLE. *Liber Algorismi de Practica Arismetrice,* in Boncompagni, Baldassarre. *Trattati d'arithmetica,* pp. 25–136. (Rome, 1857).

JORDANUS DE NEMORE. See Thomson article "Opera . . ." below for a complete listing of the works of Jordanus.

(LEONARDO FIBONACCI OF PISA). Boncompagni, Baldassarre. *Scritti di Leonardo Pisano.* 2 vols. Rome, 1857.

MARINOS OF SICHEM. (Commentary on Euclid's *Data,* above) in n. 10, pp. 1–16 and in n. 11, pp. 234–56.

PAPPUS OF ALEXANDRIA. *Pappi Alexandrini Collectionis quae supersunt.* Edited and translated into Latin by Friedrich Hultsch. 3 vols. Berlin: B. G. Teubner, 1877.

(RICHARD DE FOURNIVAL). J. J. de Vleeschauwer. *La Biblionomia de Richard de Fournival du manuscrit 636 de la Bibliothèque de la Sorbonne. Texte en facsimile avec la transcription de Léopold Delisle.* Pretoria, S. Africa: Mousain, 1965.

TRIVET, NICOLAUS. *Annales sex regum Angliae.* Edited by Thomas Hog. London, 1845.

VIÈTE, FRANÇOIS. *Introduction to the Analytic Art.* Translated by Winfree Smith, in J. Klein, *Greek Mathematical Thought and the Origin of Algebra,* pp. 313–53 (Cambridge: M.I.T. Press, 1968).

(VARIOUS). Midonick, H. P., editor. *The Treasury of Mathematics.* New York: Philosophical Library, 1965.

(VARIOUS). Thomas, Ivor. *Selections Illustrating the History of Greek Mathematics.* With an English Translation, 2 vols. Cambridge: Harvard University Press, 1957.

(VARIOUS). Thorndike, Lynn, editor. *University Records and Life in the Middle Ages.* New York: Columbia University Press, 1944.

WIDMANN, JOHANNES. *Algorithmus de Datis*, in W. Kaunzner. *Ueber Johannes Widmann von Eger*, pp. 139–42 (Munich: Deutsches Museum, 1968).

II. Secondary Sources:
Books and Monographs

AL-DAFFA', ALI ABDULLAH. *The Muslim Contribution to Mathematics.* Atlantic Highlands, NJ: Humanities Press, 1977.

ARONS, MARGUERITE. *Saint Dominic's Successor* (trans. of *Un Animateur de la jeunesse au XIII siècle*). London: Blackfriars, 1955.

BENEDICT, SUZAN ROSE. *A Comparative Study of the Early Treatises Introducing into Europe the Hindu Art of Reckoning.* Diss: University of Michigan, 1914.

BERLET, BRUNO. *Adam Riese, sein Leben, seine Rechenbücher und sein Art zu Rechen.* Leipzig, Frankfurt: Kesselringsche Hofbuchhandlung Berla, 1892.

BOYER, CARL B. *A History of Mathematics.* New York: John Wiley and Sons, 1968.

BRUINS, EVART M. and M. RUTTEN, *Textes mathématiques de Suse.* Paris: Paul Geuthner, 1961.

BUSARD, HUBERT L. L. *Quelques sujets de l'histoire des mathématiques au moyen-âge.* Paris: Palais de la Découverte, 1968.

CAJORI, FLORIAN. *A History of Mathematical Notation*, 2 vols. Chicago: Open Court Publishing Co., 1929.

CANTOR, MORITZ. *Vorlesungen über Geschichte der Mathematik*, vol. 2. New York: Johnson Reprint Corp., 1965—reprint of 2d edition, 1900.

CARRUCCIO, ETTORE. *Mathematics and Logic in History and in Contemporary Thought.* London: Faber & Faber, 1964.

CHEVALIER, U. *Répertoire des sources historiques du moyen âge, Bioliographie*, Paris: Picard, 1905.

CLAGETT, MARSHALL. *The Science of Mechanics in the Middle Ages.* Madison: University of Wisconsin Press, 1961.

COSSALI, PIETRO. *Origine dell'algebra, storia critica*, 2 vols. Parma, 1797.

CROMBIE, ARTHUR C. *Robert Grosseteste and the Origins of Experimental Science 1100–1700.* Oxford: Clarendon Press, 1953.

DELISLE, LÉOPOLD. *Le Cabinet des manuscrits de la Bibliothèque Nationale*, vol. 2. Paris, 1874.

DE MORGAN, AUGUSTUS. *Arithmetical Books from the Invention of Printing to the Present Time.* London: Taylor and Walton, 1847.

DU BOULAY (BULAEUS), CAESARE EGASSIUS. *Historia Universitatis Parisiensis*. Paris: Franciscus Noel, 1966—reprint of 1666 edition.

DUHEM, PIERRE. *Les Origines de la statique*. Paris, 1905.

HASKINS, CHARLES HOMER. *Studies in the History of Mediaeval Science*. 2d ed. New York: Frederick Ungar, 1927.

HEATH, SIR THOMAS LITTLE. *History of Greek Mathematics*, 2 vols. Oxford: Clarendon Press, 1921.

HINTIKKA, JAAKKO, AND UNTO REMES. *The Method of Analysis: Its Geometrical Origin and Its General Significance*. Boston: D. Reidel, 1974.

HOFMANN, JOSEPH E. *Geschichte der Mathematik*, vol. 1. 2d ed. Berlin: de Gruyter, 1963.

JUSCHKEWITSCH, A. P. *Geschichte der Mathematik im Mittelalter*. Leipzig: B. G. Teubner, 1964.

This is perhaps the best work in print on the history of medieval mathematics in China, India, Arabic lands, and western Europe. It deserves special attention.

KAUNZNER, WOLFGANG. *Ueber Johannes Widmann von Eger, ein Beitrag zur Geschichte der Rechenkunst in ausgehenden Mittelalter*. Quellentexte and Uebersetzungen, no. 7. Munich: Deutsches Museum, 1968.

KLEIN, JACOB. *Greek Mathematical Thought and the Origin of Algebra*. Cambridge: M.I.T. Press, 1968.

KRISTELLER, PAUL OSKAR. *Latin Manuscript Books Before 1600. A list of the Printed Catalogues and Unpublished Inventories of Extant Collections*. 3rd ed. New York: Fordham University Press, 1965.

KUHN, THOMAS S. *The Structure of Scientific Revolution*. 2d ed. Chicago: Univ. of Chicago Press, 1970.

LEFEBVRE, B. *Notes d'histoire des mathématiques (antiquité et moyen âge)*. Louvain: Société Scientifique de Bruxelles, 1920.

LINDBERG, DAVID C., editor. *Science in the Middle Ages*. Chicago: Univ. of Chicago Press, 1978.

MAAS, PAUL. *Textual Criticism*. Translated by Barbara Flower. Oxford: Clarendon Press, 1958.

MAHONEY, MICHAEL SEAN. *The Royal Road: The Development of Algebraic Analysis from 1550 to 1650, with Special Reference to the Work of Pierre de Fermat*. Diss: Princeton University, 1967.

MOODY, ERNEST A. AND MARSHALL CLAGETT, editors. *The Medieval Science of Weights*. Madison: University of Wisconsin Press, 1952.

MURRAY, ALEXANDER. *Reason and Society in the Middle Ages*. Oxford: Clarendon Press, 1978.

NESSELMANN, G. *Die Algebra der Griechen*. Berlin: G. Reimer, 1842.

NEUGEBAUER, OTTO. *The Exact Sciences in Antiquity*. 2d ed. Providence: Brown University Press, 1957.

PÉREZ, JOSÉ A. S. *La Ciencia arabe en la edad media*. Madrid: Instituto de Estudios Africanos, 1954.

RODET, LÉON. *Sur les notations numériques et Algébriques antetieurement au xvi siècle*. Paris: Ernest Leroux, 1881.

ROSE, PAUL L. *The Italian Renaissance of Mathematics*. Geneva: Librairie Droz, 1975.

SANFORD, VERA. *The History and Significance of Certain Standard Problems in Algebra*. Diss: Teacher's College, 1927.

SARTON, GEORGE. *Introduction to the History of Science*, 3 vols. Washington, D.C.: Carnegie Institution of Washington, 1927–47.

SCOTT, J. F. *A History of Mathematics*. London: Taylor & Francis, 1958.

SMITH, DAVID EUGENE. *History of Mathematics*, 2 vols, reprinted edition. New York: Dover, 1958.

———. *Rara Arithematica, A Catalogue of the Arithmetics written before the years MDCI with a Description of those in Library of George Arthur Plimpton of New York*. Boston: Ginn and Company, 1908.

STEINSCHNEIDER, MORITZ. *Die Europaischen Uebersetzungen aus dem Arabischen bis Mitte des 17. Jahrhunderts*. Graz: Akademische Druck-u. Verlagsanstalt, 1956 (reprint of journal articles of 1904–05).

THOMSON, RON B., editor. *Jordanus de Nemore and the Mathematics of Astrolabes*: *de plana spera* (An Edition with Introduction, Translation and Commentary). Toronto: Pontifical Institute of Mediaeval Studies, 1978.

TROPFKE, JOHANNES. *Geschichte der Elementar-Mathematik*, vols. I, II, III. 2d ed. Berlin: Walter de Gruyter, 1921–22.

VOSSIUS, GERHARD JOHANN. *De vniversae mathesios natvra & constitutione liber*; *cui subjungitur chronologia mathematicorvm*. Amsterdam: Joannis Blaev, 1650.

WILSON, CURTIS. *William Heytesbury, Medieval Logic and the Rise of Mathematical Physics*. Madison: University of Wisconsin Press, 1960.

ZEUTHEN, H. G. *Die Mathematik in Altertum und im Mittelalter*. Leipzig: B. G. Teubner, 1966—reprint of 1912 edition.

———. *Geschichte der Mathematik im 16. und 17. Jahrhundert*. Leipzig: B. G. Teubner, 1912.

———. *Sur l'origine de l'algèbre*. Copenhagen: Høst, 1919.

III. Secondary Sources:
Journal Articles

BONCOMPAGNI, BALDASSARRE. "Della vita et delle opere di Gherardo Cremonense," *Atti dell' Accademia de' nuovi Lincei*, 1851, *4*:412–35.

BRUINS, EVART M. "Interpretation of Cuneiform Mathematics," *Physis*, 1962, *4*:277–341.

CHASLES, MICHEL, "Histoire de l'algèbre," *Comptes Rendus*, 1841, *13*: 497–524.

CURTZE, MAXIMILIAN. "Commentar zu dem 'Tractatus *De numeris datis*' des Jordanus Neomorarius," *Zeitschrift für Mathematik und Physik, Hist-lit. Abtg.*, 1891, *36*:1–23, 41–63, 81–95, 121–38.

———. "Die Ausgabe von Jordanus' *De numeris datis* durch Prof. P. Treutlein in Karlsruhe," *Amtl. Organ der Kaiser Leop.-Carol. deutschen Akad. der Naturforscher*: Leopoldina, 1st series, 1882, *8*:26–31.

———. "Geschichte der Algebra in Deutschland im 15. Jahrhundert," *Abhandlungen Zur Geschichte der Mathematik*, 1895, *7*:33–74.

———. "Jordani Nemorarii Geometria vel de triangulis libri iv," *Mitteilungen der Coppernicusvereins für Wissenschaft und Kunst zu Thorn*, 1887, *Heft 6*.

DUTON, M. AND R. E. GRIM. "Fibonacci on Egyptian Fractions," *Fibonacci Quarterly*, 1966, *4*:339–354.

FRAJESE, ATTILIO. "L'algebra geometrica in Leonardo Pisano," *Bollettino dell'Unione Matematica Italiana*, Apr–May 1940.

GANDZ, SOLOMON. "The Origin and Development of the Quadratic Equations in Babylonian, Greek and early Arabic Algebra," *Osiris*, 1938, *3*:405–557.

———. "On the Origin of the Word 'Root'," *American Mathematical Monthly*, 1926, *33*:261–65.

———. "On the Origin of the Word 'Root'. Second Article," ibid., 1928, *35*:67–75.

———. "The Rule of Three in Arabic and Hebrew Sources," *Isis*, 1934–35, *22*:220–22.

HUGHES, BARNABAS B., O. F. M. "Biographical Information on Jordanus de Nemore to Date," *Janus*, 1975, *62*:151–56.

———. "*De Regulis Generalibus*: A Thirteenth-Century English Mathematical Tract on Problem Solving," *Viator* 1980, *11*:209–24.

———. "Johann Scheubel's Revision of Jordanus de Nemore's *De Numberis Datis*: An Analysis of an Unpublished Manuscript," *Isis*, 1972, *63*:221–234.

————. "Rhetoric, Anyone?" *The Mathematics Teacher*, 1970, *63*:267–70.

————. "Toward an Explication of Ambrosiana MS D 186 Inf." *Scriptorium*, 1972, *26*:125–27.

KARPINSKI, LOUIS C. "The Algebra of Abū Kāmil Shoja' ben Aslam," *Bibliotheca Mathematica*, 1912, ser. 3, *12*:40–55.

————. "The algebra of Abū Kāmil," *American Mathematical Monthly*, 1914, *21*:37–48.

————. "Algebraic Works to 1700," *Scripta Mathematica*, 1944, *10*:145–169.

————. "Jordanus Nemorarius and John of Halifax," *American Mathematical Monthly*, 1910, *17*:108–113.

————. "The Origin and Development of Algebra," *School Science and Mathematics*, 1923, *23*:54–64.

MAHONEY, MICHAEL SEAN. "Another Look at Greek Geometrical Analysis," *Archive for History of Exact Sciences*, 1968, *5*:318–48.

MOLLAND, A. G., "Ancestors of physics," *History of Science*, 1975, *12*:64–67.

————. "The Geometrical Background to the 'Merton School'," *British Journal for the History of Science*, 1968, *4(14)*:108–25.

RUSSEL, J. C. "Hereford and Arabic Science in England about 1175–1200," *Isis*, 1932, *18*:14–25.

SALIBA, G. A. "The Meaning of al-jabr wa'l-muqābalah," *Centaurus*, 1972–73, *17*:189–204.

SCHREIDER, N. "The Beginnings of Algebra in Mediaeval Europe as seen in the Treatise *De numeris datis* of Jordanus Nemorarius," *Ist. Math. Issled.*, 1959, *12*:679–688 (in Russian, privately translated).

SUTER, H. "Die Mathematik auf den Universitäten des Mittelalters," *Festschrift der Kantonschule in Zürich*, 1887, pp. 1–96.

THOMPSON, J. W. "The Introduction of Arabic Science into Lorraine in the Tenth Century," *Isis*, 1929, *12*:184–93.

THOMSON, RON B. "Jordanus de Nemore and the University of Toulouse," *The British Journal for the History of Science*, 1974, *7*:163–65.

————. "Jordanus de Nemore: Opera," *Mediaeval Studies*, 1976, *38*:97–144.

THOMAS, S. HARRISON. "Editing of Medieval Latin Texts in America," *Progress of Mediaeval and Renaissance Studies*, 1941, *16*:37–49.

THORNDIKE, LYNN. "Notes on Some Astronomical, Astrological and Mathematical Manuscripts of the Bibliothèque Nationale, Paris," *Journal of Warburg and Courtauld Institutes*, 1957, *20*:112–72.

TREUTLEIN, PETER. "Der Traktat des Jordanus Nemorarius *De numeris datis*," *Abhandlungen zur Geschichte der Mathematik*, 1879, 2:127–66.

———. "Die deutsche Coss," *Zeitschrift für Mathematik, Astronomie und Physik, Hist.-lit. Abtg.*, 1879, 24:1–124.

VON STERNECK, R. DAUBLEBSKY. "Zur vervollständigung der Ausgaben des Schrift des Jordanus Nemorarius: 'Tractatus de Numeris Datis'," *Monatshefte für Mathematik und Physik*, 1896, 7:165–79.

WAPPLER, E. "Zur Geschichte der deutschen Algebra im 15. Jahrhundert," *Program*, Zwickau, 1887, pp. 539–54.

———. "Zur Geschichte der Mathematik," *Zeitschrift für Mathematik, Astronomie und Physik, Hist.-lit. Abtg.*, 1900, 45:7–9.

WERTHEIM, G. "Uber die Lösung einiger Aufgaben im *Tractatus de numeris datis* des Jordanus Nemorarius," *Bibliotheca mathematica*, 3rd series, 1900, 1:417–20.

Index of Latin Terms

Indexed here are most of the Latin words found in the text. Omitted from the list are forms of the verbs *facere*, *fieri*, and *esse*, all prepositions, conjunctions, and common comparative terms (like *maior* and *minor*), together with the integers and fractions. I have not gathered every location of a particular word; for most of them I found two citations. Finally, regardless of how the word appears in the text I have adjusted all verbs into the present infinitive, nouns into the nominative singular, and adjectives into the masculine nominative singular. Again, the Roman numeral refers to the book, the Arabic to the proposition.

Index

The bulk of the index identifies proper names that may be distinguished into two groups. First are the names of those who influenced the development of mathematics. Second are the names of authors, editors, or translators whose works appear in the contents of the various codices containing copies of *De datis*. The respective works of the latter group are not included in the index, to avoid needless repetition of titles easily found in the text. Following each proper name where possible, reference is made to *Dictionary of Scientific Biography* (DSB); additional biographical and bibliographic information may be consulted there. The many repetitious references in the footnotes to possible sources of, or to theorem similar to, propositions in *De datis* are not recorded below. Nor has the bibliography been indexed. The remainder of the index houses items of mathematical interest.

Designer: Wolfgang Lederer
Compositor: Syntax International

Text: Times Roman
Display: Times Roman